OFFICIAL
SQA SPECIMEN
QUESTION PAPER
AND HODDER GIBSON
MODEL QUESTION PAPERS
WITH ANSWERS

NATIONAL 5

GEOGRAPHY

2013 Specimen Question Paper & 2013 Model Papers

HODDER
GIBSON
LEARN MORE

This book contains the official 2013 SQA Specimen Question Paper for National 5 Geography, with associated SQA approved answers modified from the official marking instructions that accompany the paper.

In addition the book contains model practice papers, together with answers, plus study skills advice. These papers, some of which may include a limited number of previously published SQA questions, have been specially commissioned by Hodder Gibson, and have been written by experienced senior teachers and examiners in line with the new National 5 syllabus and assessment outlines, Spring 2013. This is not SQA material but has been devised to provide further practice for National 5 examinations in 2014 and beyond.

Hodder Gibson is grateful to the copyright holders, as credited on the final page of the Answer Section, for permission to use their material. Every effort has been made to trace the copyright holders and to obtain their permission for the use of copyright material. Hodder Gibson will be happy to receive information allowing us to rectify any error or omission in future editions.

Hachette UK's policy is to use papers that are natural, renewable and recyclable products and made from wood grown in sustainable forests. The logging and manufacturing processes are expected to conform to the environmental regulations of the country of origin.

Orders: please contact Bookpoint Ltd, 130 Park Drive, Abingdon, Oxon OX14 4SE. Telephone: (44) 01235 827720. Fax: (44) 01235 400454. Lines are open 9.00–5.00, Monday to Saturday, with a 24-hour message answering service. Visit our website at www.hoddereducation.co.uk. Hodder Gibson can be contacted direct on: Tel: 0141 848 1609; Fax: 0141 889 6315; email: hoddergibson@hodder.co.uk

This collection first published in 2013 by
Hodder Gibson, an imprint of Hodder Education,
An Hachette UK Company
2a Christie Street
Paisley PA1 1NB

BrightRED   Hodder Gibson is grateful to Bright Red Publishing Ltd for collaborative work in preparation of this book and all SQA Past Paper and National 5 Model Paper titles 2013.

Specimen Question Paper © Scottish Qualifications Authority. Answers, Model Question Papers, and Study Skills Section © Hodder Gibson. Model Question Papers creation/compilation, Answers and Study Skills section © Dr William Dick and Sheena Williamson. All rights reserved. Apart from any use permitted under UK copyright law, no part of this publication may be reproduced or transmitted in any form or by any means, electronic or mechanical, including photocopying and recording, or held within any information storage and retrieval system, without permission in writing from the publisher or under licence from the Copyright Licensing Agency Limited. Further details of such licences (for reprographic reproduction) may be obtained from the Copyright Licensing Agency Limited, Saffron House, 6–10 Kirby Street, London EC1N 8TS.

Typeset by PDQ Digital Media Solutions Ltd, Bungay, Suffolk NR35 1BY

Printed in the UK

A catalogue record for this title is available from the British Library

ISBN: 978-1-4718-0211-9

3 2 1

2014 2013

Introduction

Study Skills – what you need to know to pass exams!

Pause for thought

Many students might skip quickly through a page like this. After all, we all know how to revise. Do you really though?

Think about this:

"IF YOU ALWAYS DO WHAT YOU ALWAYS DO, YOU WILL ALWAYS GET WHAT YOU HAVE ALWAYS GOT."

Do you like the grades you get? Do you want to do better? If you get full marks in your assessment, then that's great! Change nothing! This section is just to help you get that little bit better than you already are.

There are two main parts to the advice on offer here. The first part highlights fairly obvious things but which are also very important. The second part makes suggestions about revision that you might not have thought about but which WILL help you.

Part 1

DOH! It's so obvious but …

Start revising in good time

Don't leave it until the last minute – this will make you panic.

Make a revision timetable that sets out work time AND play time.

Sleep and eat!

Obvious really, and very helpful. Avoid arguments or stressful things too – even games that wind you up. You need to be fit, awake and focused!

Know your place!

Make sure you know exactly **WHEN and WHERE** your exams are.

Know your enemy!

Make sure you know what to expect in the exam.

How is the paper structured?

How much time is there for each question?

What types of question are involved?

Which topics seem to come up time and time again?

Which topics are your strongest and which are your weakest?

Are all topics compulsory or are there choices?

Learn by DOING!

There is no substitute for past papers and practice papers – they are simply essential! Tackling this collection of papers and answers is exactly the right thing to be doing as your exams approach.

Part 2

People learn in different ways. Some like low light, some bright. Some like early morning, some like evening / night. Some prefer warm, some prefer cold. But everyone uses their BRAIN and the brain works when it is active. Passive learning – sitting gazing at notes – is the most INEFFICIENT way to learn anything. Below you will find tips and ideas for making your revision more effective and maybe even more enjoyable. What follows gets your brain active, and active learning works!

Activity 1 – Stop and review

Step 1

When you have done no more than 5 minutes of revision reading STOP!

Step 2

Write a heading in your own words which sums up the topic you have been revising.

Step 3

Write a summary of what you have revised in no more than two sentences. Don't fool yourself by saying, 'I know it but I cannot put it into words'. That just means you don't know it well enough. If you cannot write your summary, revise that section again, knowing that you must write a summary at the end of it. Many of you will have notebooks full of blue/black ink writing. Many of the pages will not be especially attractive or memorable so try to liven them up a bit with colour as you are reviewing and rewriting. **This is a great memory aid, and memory is the most important thing.**

Activity 2 — Use technology!

Why should everything be written down? Have you thought about 'mental' maps, diagrams, cartoons and colour to help you learn? And rather than write down notes, why not record your revision material?

What about having a text message revision session with friends? Keep in touch with them to find out how and what they are revising and share ideas and questions.

Why not make a video diary where you tell the camera what you are doing, what you think you have learned and what you still have to do? No one has to see or hear it but the process of having to organise your thoughts in a formal way to explain something is a very important learning practice.

Be sure to make use of electronic files. You could begin to summarise your class notes. Your typing might be slow but it will get faster and the typed notes will be easier to read than the scribbles in your class notes. Try to add different fonts and colours to make your work stand out. You can easily Google relevant pictures, cartoons and diagrams which you can copy and paste to make your work more attractive and **MEMORABLE**.

Activity 3 – This is it. Do this and you will know lots!

Step 1

In this task you must be very honest with yourself! Find the SQA syllabus for your subject (www.sqa.org.uk). Look at how it is broken down into main topics called MANDATORY knowledge. That means stuff you MUST know.

Step 2

BEFORE you do ANY revision on this topic, write a list of everything that you already know about the subject. It might be quite a long list but you only need to write it once. It shows you all the information that is already in your long-term memory so you know what parts you do not need to revise!

Step 3

Pick a chapter or section from your book or revision notes. Choose a fairly large section or a whole chapter to get the most out of this activity.

With a buddy, use Skype, Facetime, Twitter or any other communication you have, to play the game "If this is the answer, what is the question?". For example, if you are revising Geography and the answer you provide is "meander", your buddy would have to make up a question like "What is the word that describes a feature of a river where it flows slowly and bends often from side to side?".

Make up 10 "answers" based on the content of the chapter or section you are using. Give this to your buddy to solve while you solve theirs.

Step 4

Construct a wordsearch of at least 10 X 10 squares. You can make it as big as you like but keep it realistic. Work together with a group of friends. Many apps allow you to make wordsearch puzzles online. The words and phrases can go in any direction and phrases can be split. Your puzzle must only contain facts linked to the topic you are revising. Your task is to find 10 bits of information to hide in your puzzle but you must not repeat information that you used in Step 3. DO NOT show where the words are. Fill up empty squares with random letters. Remember to keep a note of where your answers are hidden but do not show your friends. When you have a complete puzzle, exchange it with a friend to solve each other's puzzle.

Step 5

Now make up 10 questions (not "answers" this time) based on the same chapter used in the previous two tasks. Again, you must find NEW information that you have not yet used. Now it's getting hard to find that new information! Again, give your questions to a friend to answer.

Step 6

As you have been doing the puzzles, your brain has been actively searching for new information. Now write a NEW LIST that contains only the new information you have discovered when doing the puzzles. Your new list is the one to look at repeatedly for short bursts over the next few days. Try to remember more and more of it without looking at it. After a few days, you should be able to add words from your second list to your first list as you increase the information in your long-term memory.

FINALLY! Be inspired...

Make a list of different revision ideas and beside each one write **THINGS I HAVE** tried, **THINGS I WILL** try and **THINGS I MIGHT** try. Don't be scared of trying something new.

And remember – "FAIL TO PREPARE AND PREPARE TO FAIL!"

Introduction
National 5 Geography

The exam

The course assessment will consist of two components: a question paper and an assignment.

The question paper

The purpose of this question paper is to assess your application of skills, and breadth of knowledge and understanding across the three units of the course.

This question paper will give you an opportunity to demonstrate the following higher-order cognitive skills and knowledge and understanding from the mandatory content of the course:

- using a limited range of mapping skills;
- using a limited range of numerical and graphical information;
- giving detailed descriptions and explanations with some analysis.

The question paper will have 60 marks (75% of the total mark) distributed across three sections.

Section 1: Physical Environments

This section will have 20 marks and will be made up of limited/extended-response questions. These require you to draw on your knowledge and understanding, and to apply the skills you have acquired during the course.

Section 2: Human Environments

This section will have 20 marks and will be made up of limited/extended-response questions. These require you to draw on your knowledge and understanding, and to apply the skills you have acquired during the course.

These questions will draw on the knowledge and understanding and skills described in the 'further mandatory information on course coverage' section.

Section 3: Global Issues

This section will have 20 marks and will be made up of limited/extended-response questions. These require you to draw on your knowledge and understanding, and to apply the skills you have acquired during the course.

These questions will draw on the knowledge and understanding and skills described in the 'further mandatory information on course coverage' section. In this section you will be required to attempt two questions from six. The choice of topics are: Climate Change; Impact of Human Activity on the Natural Environment; Environmental Hazards; Trade and Globalisation; Tourism; and Health.

What you will be tested on

For marks to be given, points must relate to the question asked.

There are six types of question used in this paper:

A. Describe	D. Match
B. Explain	E. Give map evidence
C. Give reasons	F. Give advantages and/or disadvantages

Questions which ask candidates to **describe**:

You must make a number of relevant, factual points. These should be key points taken from a given source, for example a map, diagram or table.

Questions which ask candidates to **explain** or **give reasons**:

You should make a number of points giving clear reasons for a given situation. The command word 'explain' will be used when you are asked to demonstrate knowledge and understanding. Sometimes the command words 'give reasons' may be used as an alternative to 'explain'.

Questions which ask candidates to **match**:

You are asked to match two sets of variables, for example to match features to a correct grid reference.

Questions which ask candidates to **give map evidence**:

You should look for evidence on the map and make clear statements to support your answer.

Questions which ask candidates to **give advantages and/or disadvantages**:

You should select relevant advantages or disadvantages of a proposed development, for example the location of a new shopping centre, and demonstrate your understanding of the significance of the proposal.

Some tips for revising

- To be best prepared for the examination, organise your notes into sections. Try to work out a schedule for studying with a programme which includes the sections of the syllabus you intend to study.
- Organise your notes into checklists and revision cards.
- Try to avoid leaving your studying to a day or two before the exam. Also try to avoid cramming your studies into the night before the examination, and especially avoid staying up late to study.
- One useful technique when revising is to use summary note cards on individual topics.

- Make use of specimen and model paper questions to test your knowledge or enquiry skills. Go over your answers and give yourself a mark for every correct point you make when comparing your answer with your notes.
- If you work with a classmate, try to mark each other's practice answers.
- Practise your diagram-drawing skills and your writing skills. Ensure that your answers are clearly worded. Try to develop the points that you make in your answers.

Some tips for the exam

- Do not write lists, even if you are running out of time. You will lose marks. If the question asks for an opinion based on a choice, for example on the suitability of a particular site or area for a development, do not be afraid to refer to negative points such as why the alternatives are not as good. You will get credit for this.
- Make sure you have a copy of the examination timetable and have planned a schedule for studying.
- Arrive at the examination in plenty of time with the appropriate equipment – pen, pencil, rubber and ruler.
- Carefully read the instructions on the paper and at the beginning of each part of the question.
- Answer all of the compulsory questions in each paper you sit.
- Use the number of marks as a guide to the length of your answer.
- Try to include examples in your answer wherever possible. If asked for diagrams, draw clear, labelled diagrams.
- Read the question instructions very carefully. If the question asks you to 'describe', make sure that this is what you do.
- If you are asked to 'explain', you must use phrases such as 'due to', 'this happens because' and 'this is a result of'. If you describe rather than explain, you will lose most of the marks for that question.
- If you finish early, do not leave the exam. Use the remaining time to check your answers and go over any questions which you have partially answered, especially Ordnance Survey map questions.
- Practise drawing diagrams which may be included in your answers, for example corries or pyramidal peaks.
- Make sure that you have read the instructions on the question carefully and that you have avoided needless errors. For example, answering the wrong sections or failing to explain when asked to, or perhaps omitting to refer to a named area or case study.

- One technique which you might find helpful, especially when answering long questions worth 10 or more marks, is to 'brainstorm' possible points for your answer. You can write these down in a list at the start of your answer. As you go through your answer, you can double-check with your list to ensure that you have put as much into your answer as you can. This stops you from coming out of the exam and being annoyed that you forgot to mention an important point.

Common errors

Markers of the external examination often remark on errors which occur frequently in candidates' answers. These include the following:

Lack of sufficient detail

- Many candidates fail to provide sufficient detail in answers, often by omitting reference to specific examples, or not elaborating or developing points made in their answer. As noted above, a good guide to the amount of detail required is the number of marks given for the question. If, for example, the total marks offered is 6, then you should make at least six valid points.

Listing

- If you write a simple list of points rather than fuller statements in your answer, you will automatically lose marks. For example, in a 4 mark question, you will obtain only 1 mark for a list.
- The same rule applies to a simple list of bullet points. However, if you couple bullet points with some detailed explanation, you could achieve full marks.

Irrelevant answers

- You must read the question instructions carefully so as to avoid giving answers which are irrelevant to the question. For example, if you are asked to 'explain' and you simply 'describe', you will lose marks. If you are asked for a named example and you do not provide one, you will forfeit marks.

Repetition

- You should be careful not to repeat points already made in your answer. These will not gain any further marks. You may feel that you have written a long answer, but it may contain the same basic information repeated again and again. Unfortunately, these repeated statements will be ignored by the marker.

Good luck!

Remember that the rewards for passing National 5 Geography are well worth it! Your pass will help you to get the future you want for yourself. In the exam, be confident in your own ability. If you're not sure how to answer a question, trust your instincts and just give it a go anyway. Keep calm and don't panic! GOOD LUCK!

NATIONAL 5

2013 Specimen Question Paper

HODDER
GIBSON
LEARN MORE

FOR OFFICIAL USE

Mark

National Qualifications SPECIMEN ONLY

SQ19/N5/01

Geography

Date – N/A for specimen

Duration – 1 hour and 30 minutes

Fill in these boxes and read what is printed below.

Full name of centre

Town

Forename(s)

Surname

Number of seat

Date of birth

Day	Month	Year
D D	M M	Y Y

Scottish candidate number

Total marks — 60

SECTION 1 — PHYSICAL ENVIRONMENTS — 20 marks

Attempt EITHER question 1 **or** question 2 AND questions 3, 4 and 5

SECTION 2 — HUMAN ENVIRONMENTS — 20 marks

Attempt questions 6, 7 and 8

SECTION 3 — GLOBAL ISSUES — 20 marks

Attempt any TWO of the following

Question 9 — Climate Change

Question 10 — Impact of Human Activity on the Natural Environment

Question 11 — Environmental Hazards

Question 12 — Trade and Globalisation

Question 13 — Tourism

Question 14 — Health

Before attempting the questions you must check that your answer booklet is for the same subject and level as the question paper.

You should read the questions carefully.

On the answer booklet, you must clearly identify the question number you are attempting.

Credit will always be given for appropriately labelled sketch maps and diagrams.

Use **blue** or **black** ink.

Before leaving the examination room you must give your answer booklet to the Invigilator. If you do not, you may lose all the marks for this paper.

OS MAP ITEM A

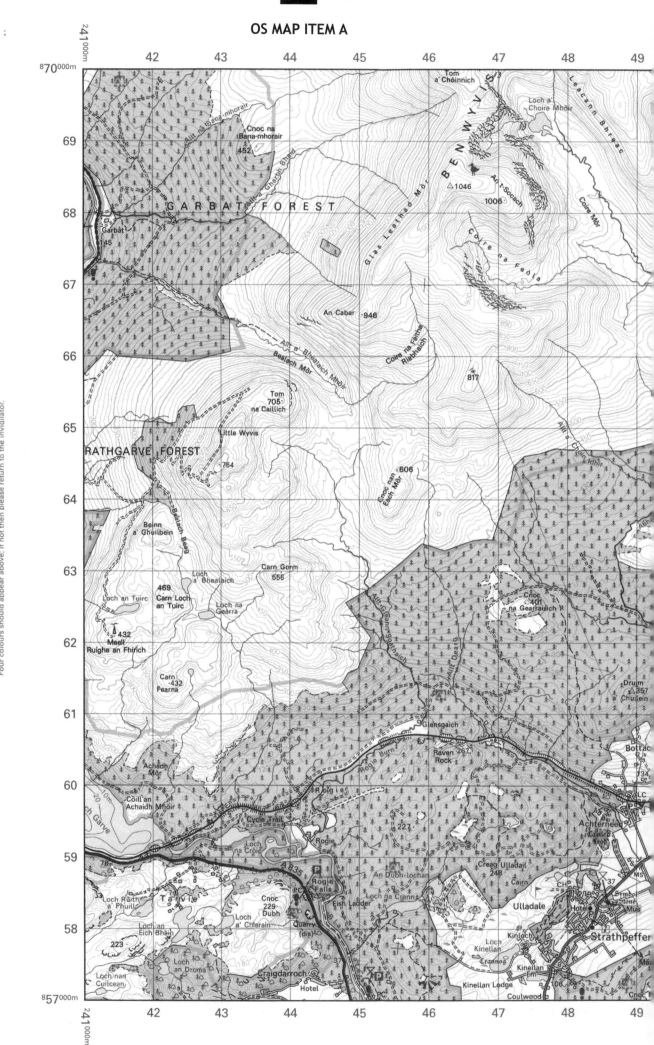

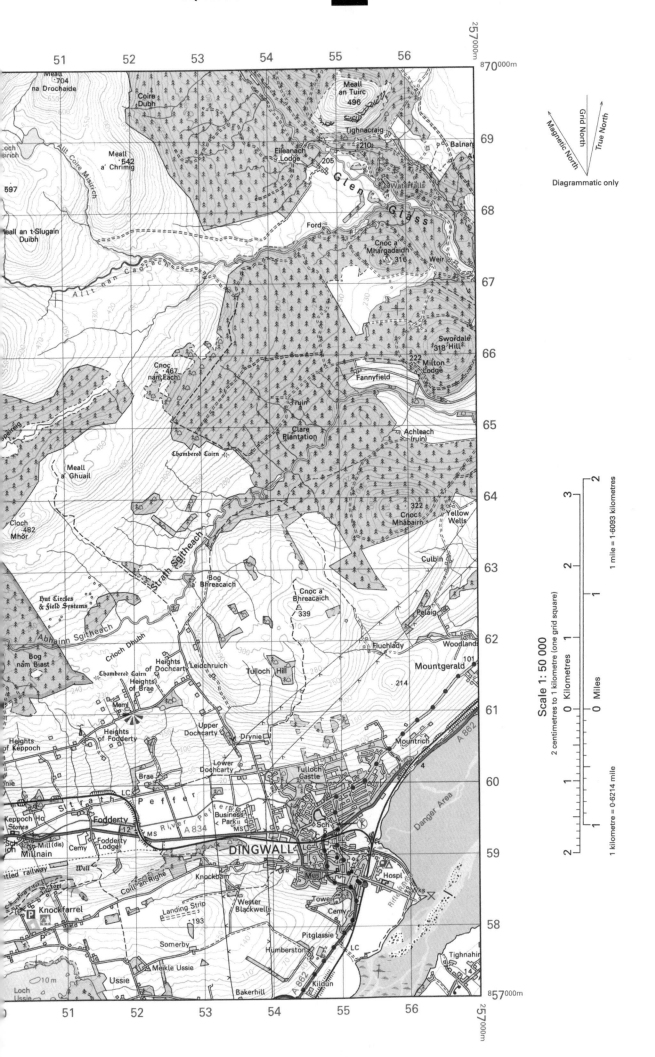

Scale 1:50 000

2 centimetres to 1 kilometre (one grid square)

1 mile = 1·6093 kilometres

1 kilometre = 0·6214 mile

MARKS

SECTION 1 — PHYSICAL ENVIRONMENTS — 20 marks

Attempt EITHER Question 1 or Question 2
AND Questions 3, 4 and 5

Question 1 — Glaciated Uplands

(a) Study OS map **Item A** of the Dingwall area.

Use the information in the OS map **Item A** to **match** the features of glaciated uplands in the table below with the correct grid reference.

Features of glaciated uplands		
U-shaped valley	corrie	truncated spur

Choose from grid references			
525658	467677	483685	476683

3

(b) **Explain** the formation of a U-shaped valley.

You may use a diagram(s) in your answer.

4

Total marks   7

NOW ATTEMPT QUESTIONS 3, 4 AND 5

MARKS | DO NOT WRITE IN THIS MARGIN

DO NOT ATTEMPT THIS QUESTION IF YOU HAVE ALREADY ANSWERED QUESTION 1

Question 2 — Rivers and Valleys

(a) Study OS map **Item A** of the Dingwall area.

Use the information in the OS map Item A to **match** the features of rivers and valleys in the table below with the correct grid reference.

Features of rivers and valleys		
V-shaped valley	meander	river flowing NW

Choose from grid references			
522623	457668	523594	435663

3

(b) **Explain** the formation of an ox-bow lake.

You may wish to use a diagram(s) in your answer.

4

Total marks **7**

NOW ATTEMPT QUESTIONS 3, 4 AND 5

MARKS | DO NOT WRITE IN THIS MARGIN

NOW ATTEMPT QUESTIONS 3, 4 AND 5

Question 3

Diagram Q3A — Cross-section from GR 466658 to GR 510580

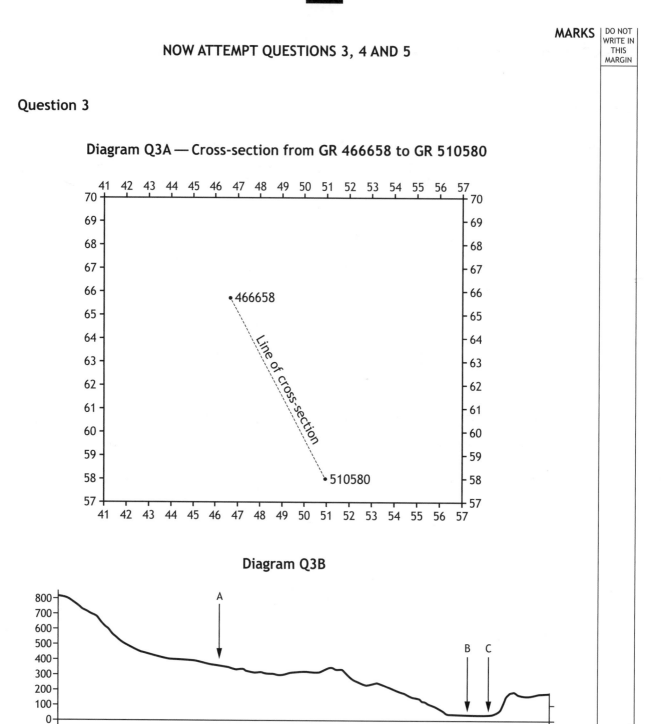

Diagram Q3B

Study OS map **Item A** of the Dingwall area and find the cross-section shown on Diagrams Q3A and Q3B above.

Use the information in the OS map Item A to **match** the letters A to C with the correct features, choosing from the features below.

railway	forestry	River Peffer	dismantled railway

3

MARKS | DO NOT WRITE IN THIS MARGIN

Question 4

Diagram Q4 — Synoptic Chart for 15 January 2010

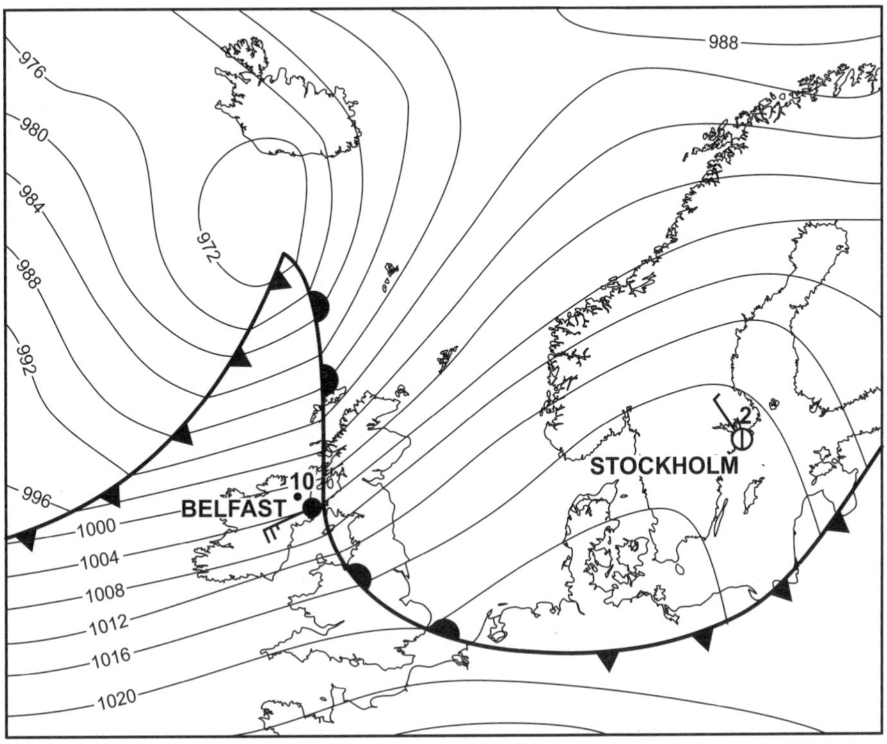

Belfast

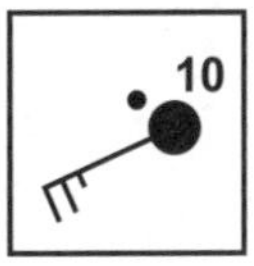

Stockholm

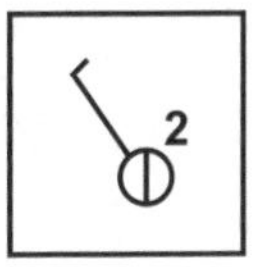

Study Diagram Q4 above.

Use the information in Diagram Q4 to **give reasons** for the differences in the weather conditions between Belfast and Stockholm. 4

OS MAP ITEM B

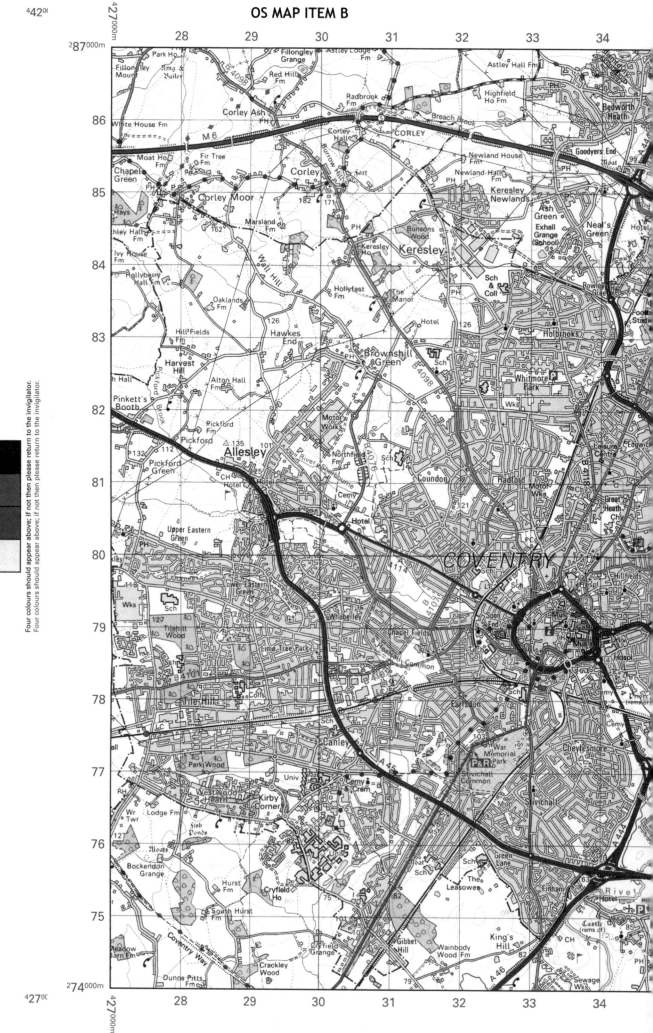

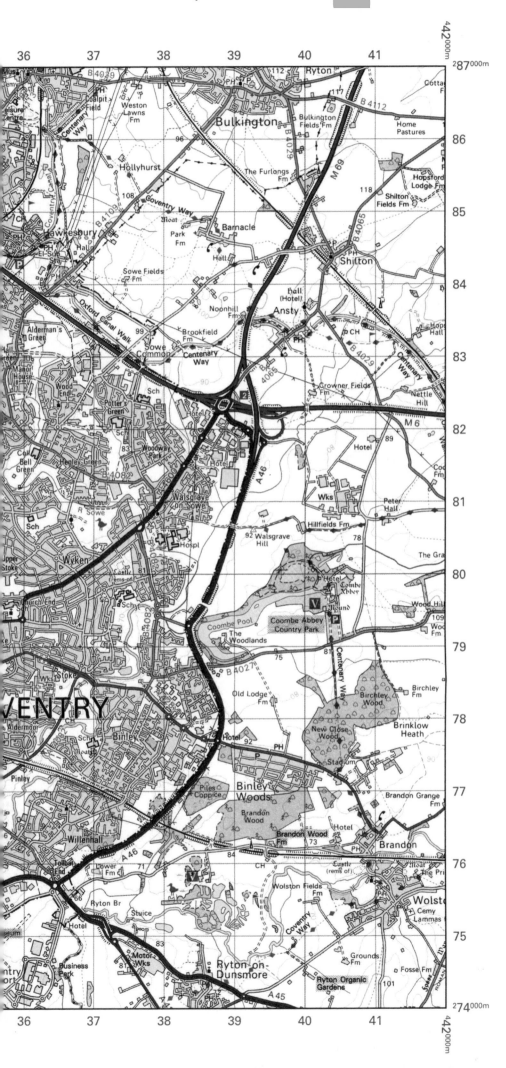

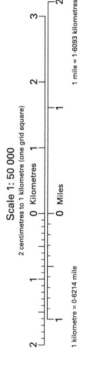

Scale 1 : 50 000

2 centimetres to 1 kilometre (one grid square)

1 mile = 1·6093 kilometres

1 kilometre = 0·6214 mile

Grid North
Magnetic North
True North

Diagrammatic only

Question 5

Diagram Q5 — Land Uses in Different Landscapes

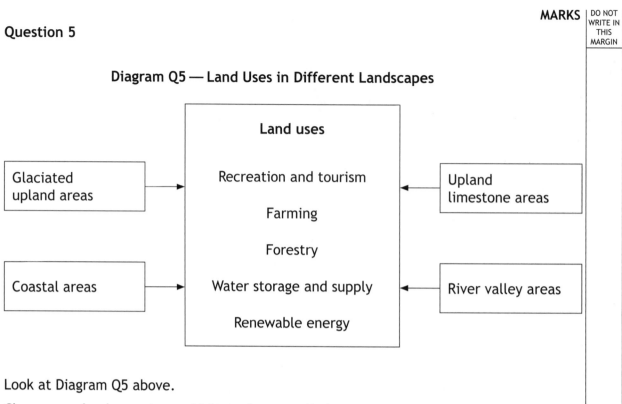

Look at Diagram Q5 above.

Choose **one** landscape type which you have studied.

For the landscape type you have chosen **describe, in detail,** ways in which at least **two** of the land uses shown can be in conflict.

6

MARKS | DO NOT WRITE IN THIS MARGIN

SECTION 2 — HUMAN ENVIRONMENTS — 20 marks

Attempt Questions 6, 7 and 8

Question 6

Study OS map **Item B** of the Coventry area.

(a) **Use Item B to give map evidence** which shows that part of Coventry's CBD is located in grid square 3379.   **3**

(b) There is a plan to build a shopping centre in grid square 3981.

Give advantages **and/or** disadvantages of this location for a shopping centre. You **must** use map evidence from Item B to support your answer.   **5**

Total marks   **8**

Question 7

Diagram Q7 — Modern factors affecting farming

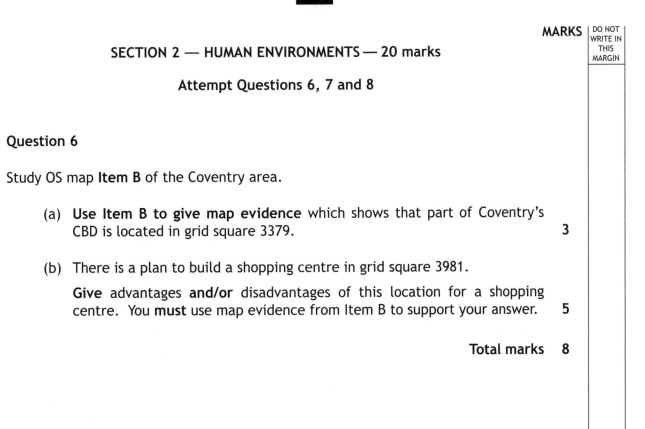

Organic farming	Diversification	GM crops
Biofuels	New technology	Government policy

Look at Diagram Q7.

For **either** a developed **or** developing country, **describe in detail** the effects of **two** of the factors shown in Diagram Q7.   **6**

Question 8

Diagram Q8 — Population Pyramids for Kenya and the United States

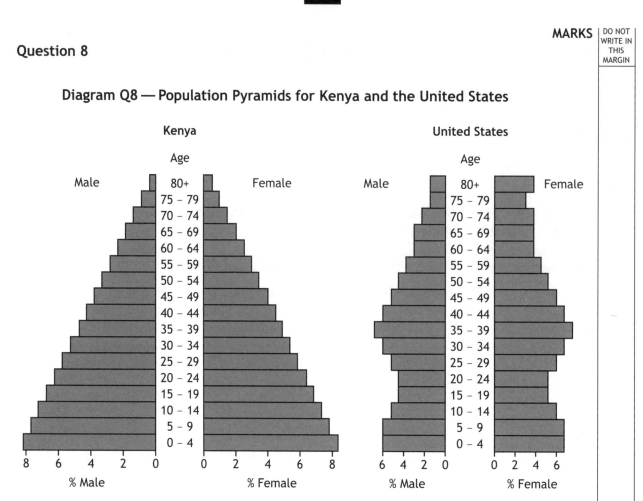

Look at Diagram Q8.

Explain the differences between the population structures of Kenya and the United States.

6

NOW GO TO SECTION 3

MARKS | DO NOT WRITE IN THIS MARGIN

SECTION 3 — GLOBAL ISSUES — 20 marks

Attempt any TWO questions

Question   9 — Climate Change

Question 10 — Impact of Human Activity on the Natural Environment

Question 11 — Environmental Hazards

Question 12 — Trade and Globalisation

Question 13 — Tourism

Question 14 — Health

MARKS | DO NOT WRITE IN THIS MARGIN

Question 9 — Climate Change

Diagram Q9 — Average Global Temperature (degrees C)

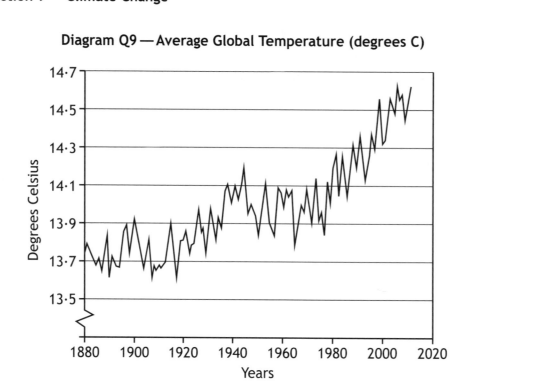

(a) Study Diagram Q9.

Use the information in Diagram Q9 to **describe, in detail,** the pattern of average global temperature over time. **4**

(b) **Explain, in detail**, the likely effects of climate change on people **and** the environment.

Your answer should refer to examples you have studied. **6**

Total marks 10

MARKS | DO NOT WRITE IN THIS MARGIN

Question 10 — Impact of Human Activity on the Natural Environment

Diagram Q10 — Climate Graphs

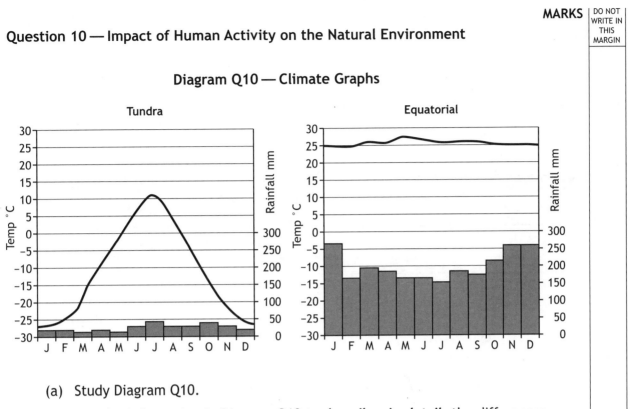

(a) Study Diagram Q10.

Use the information in Diagram Q10 to **describe, in detail,** the differences between the two climates shown.   **4**

(b) **Explain, in detail,** the impact of human activity on people **and** the environment.

Your answer should refer to examples you have studied.   **6**

Total marks **10**

MARKS

Question 11 — Environmental Hazards

Diagram Q11 — Tropical Storms

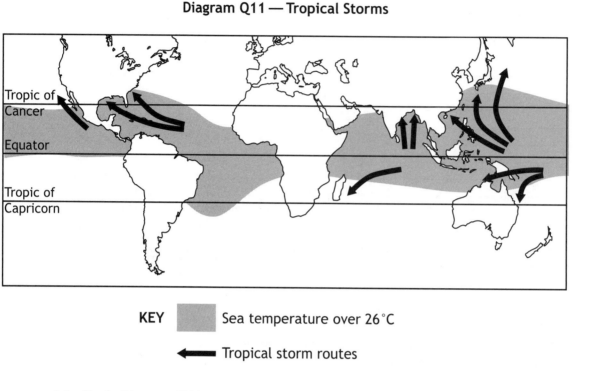

KEY   Sea temperature over 26°C

    ◀—— Tropical storm routes

(a) Study Diagram Q11.

Use the information in Diagram Q11 to **describe, in detail,** where tropical storms are found throughout the world.

4

(b) **Explain, in detail,** the ways in which aid can lessen the impact of environmental hazards.

Your answer should refer to examples you have studied.

6

Total marks 10

Question 12 — Trade and Globalisation

Diagram Q12 — Main Imports and Exports for a Selected Country

Main Exports	%	Main Imports	%
Coffee, tea, spices	19	Fuel, including oil	24
Timber	10	Machinery	9
Vegetables	5	Electrical and electronic equipment	7
Minerals and metals	4	Vehicles	7
Total value	*$4·8 billion*	*Total value*	*$11 billion*

(a) Study Diagram Q12.

Do these figures show a **developed** or **developing** country?

Use the information in Diagram Q12 to **give detailed reasons** for your answer.

4

(b) **Explain**, **in detail**, the ways in which Fair Trade helps farmers in the Developing World.

Your answer should refer to examples you have studied.

6

Total marks 10

MARKS DO NOT WRITE IN THIS MARGIN

Question 13 — Tourism

Diagram Q13 — International Tourist Destination Numbers (millions)

Destinations	1995	2010	2020 (forecast)
Africa	20	47	77
Americas	109	190	262
East Asia/Pacific	81	195	397
Europe	338	527	717
Middle East	12	36	69
South Asia	4	11	19
Total	**564**	**1,006**	**1,541**
Long-haul travel	101 (18%)	216 (21%)	378 (25%)

(a) Study Diagram Q13.

Use the information in Diagram Q13 to **describe, in detail,** the changes in international tourist destination numbers between 1995 and 2020. 4

(b) **Explain, in detail,** the benefits and problems caused by increasing tourism in developing countries.

Your answer should refer to examples you have studied. 6

Total marks 10

Question 14 — Health

Diagram Q14: Average Life Expectancy (years)

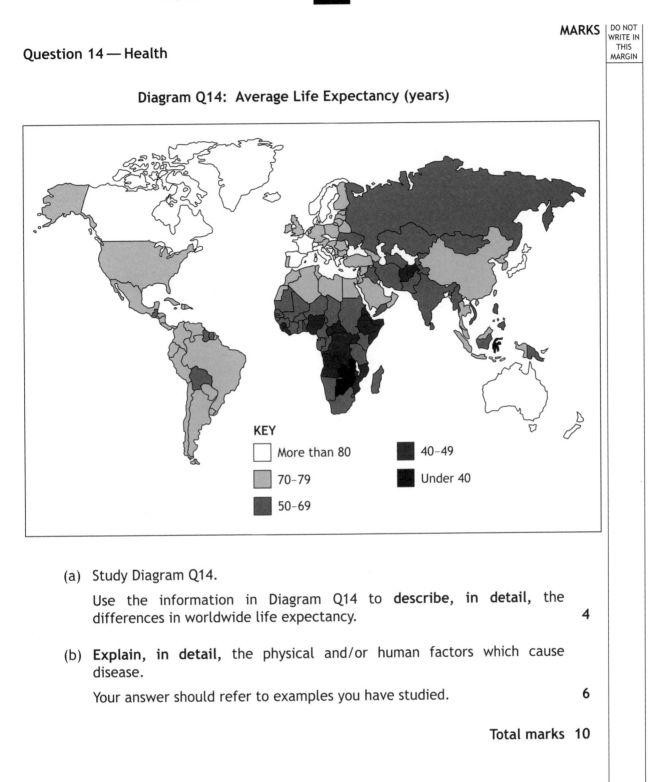

KEY

- More than 80
- 70–79
- 50–69
- 40–49
- Under 40

(a) Study Diagram Q14.

Use the information in Diagram Q14 to **describe, in detail,** the differences in worldwide life expectancy.　　　　**4**

(b) **Explain, in detail,** the physical and/or human factors which cause disease.

Your answer should refer to examples you have studied.　　　　**6**

Total marks　**10**

[END OF SPECIMEN PAPER]

NATIONAL 5

2013 Model Paper 1

National Qualifications
MODEL PAPER 1

Geography

Duration — 1 hour and 30 minutes

Total marks — 60

SECTION 1 — PHYSICAL ENVIRONMENTS — 20 marks

Attempt EITHER question 1 **or** question 2 AND questions 3, 4 and 5

SECTION 2 — HUMAN ENVIRONMENTS — 20 marks

Attempt questions 6, 7 and 8

SECTION 3 — GLOBAL ISSUES — 20 marks

Attempt any TWO of the following

Question   9 — Climate Change

Question 10 — Impact of Human Activity on the Natural Environment

Question 11 — Environmental Hazards

Question 12 — Trade and Globalisation

Question 13 — Tourism

Question 14 — Health

You should read the questions carefully.

Credit will always be given for appropriately labelled sketch maps and diagrams.

Use **blue** or **black** ink.

OS MAP ITEM A

Extract No 1655/36

1:50 000 Scale
Landranger Series

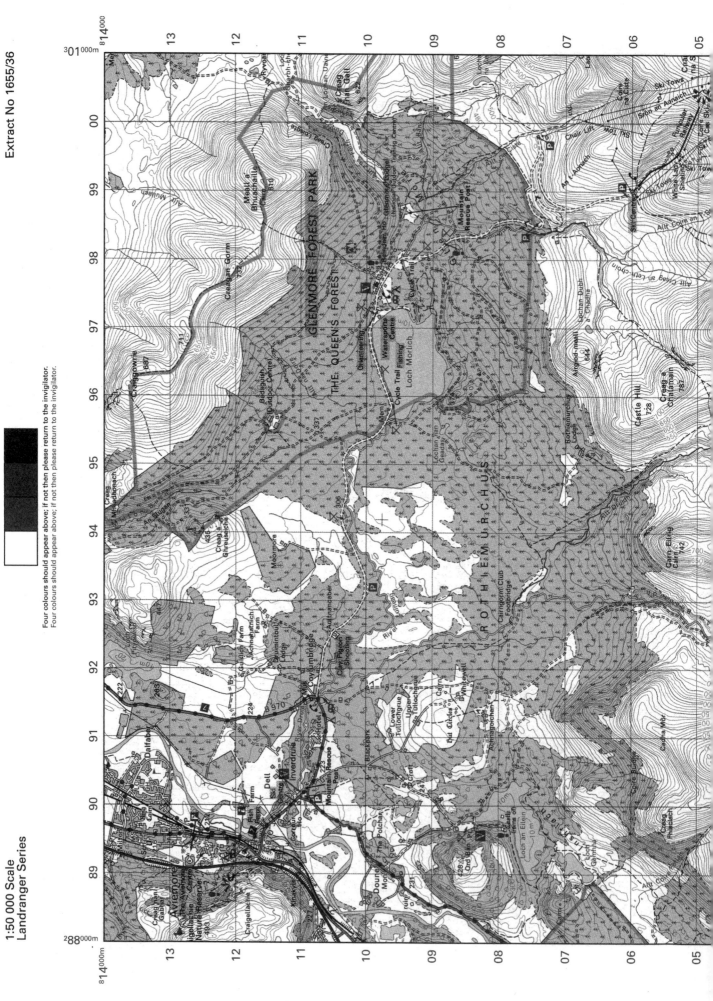

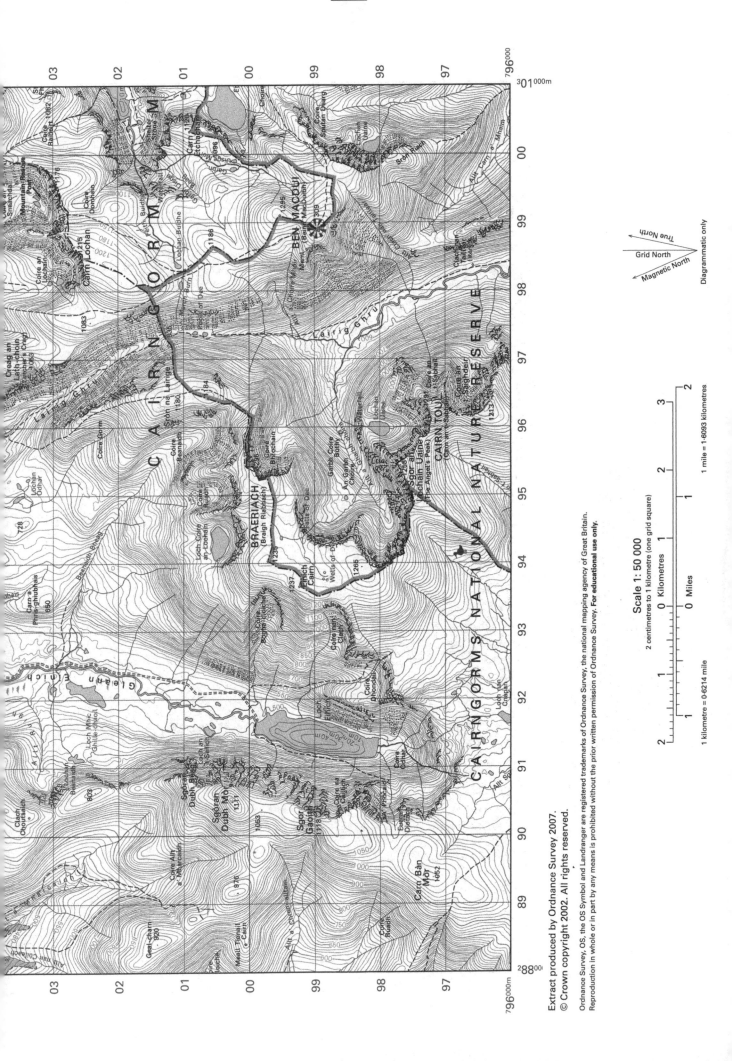

Extract produced by Ordnance Survey 2007.
© Crown copyright 2002. All rights reserved.

Scale 1: 50 000

2 centimetres to 1 kilometre (one grid square)

1 mile = 1·6093 kilometres

1 kilometre = 0·6214 mile

OS MAP ITEM B

Extract No 1939/OL12

1:25 000 Scale
Explorer Series

Four colours should appear above; if not then please return to the invigilator.
Four colours should appear above; if not then please return to the invigilator.

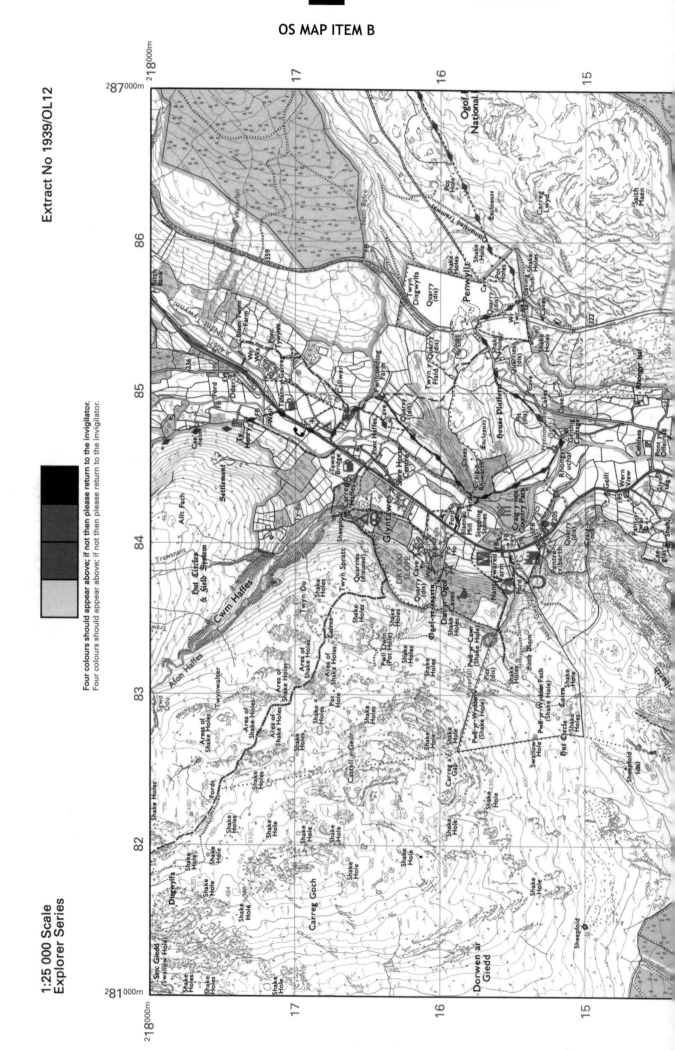

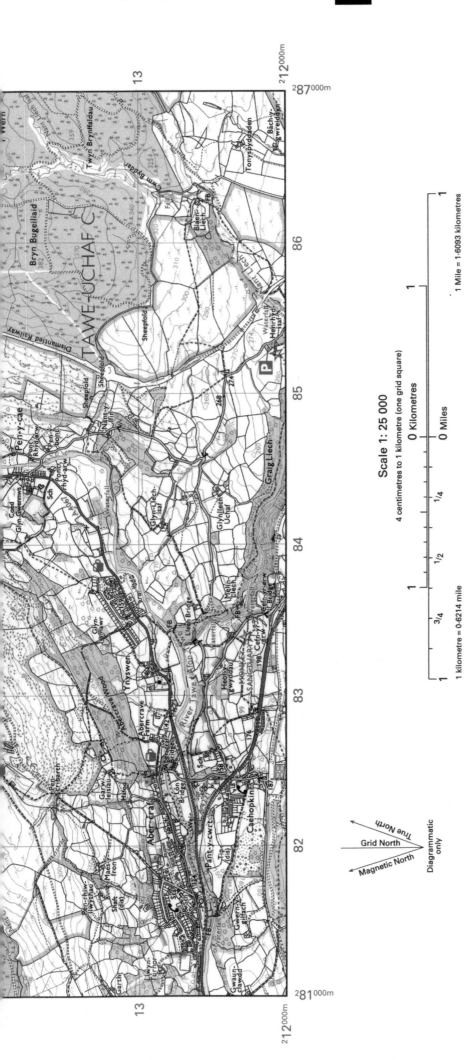

Scale 1: 25 000

4 centimetres to 1 kilometre (one grid square)

0 Kilometres

0 Miles

1 kilometre = 0·6214 mile

1 Mile = 1·6093 kilometres

Magnetic North

Grid North

True North

Diagrammatic only

Extract produced by Ordnance Survey 2011. Licence: 100035658
© Crown copyright 2008. All rights reserved

Ordnance Survey, OS, the OS Symbol and Explorer are registered trademarks of Ordnance Survey, the national mapping agency of Great Britain.
Reproduction in whole or in part by any means is prohibited without the prior written permission of Ordnance Survey. **For educational use only.**

MARKS | DO NOT WRITE IN THIS MARGIN

SECTION 1 — PHYSICAL ENVIRONMENTS — 20 marks

Attempt EITHER Question 1 or Question 2
AND Questions 3, 4 and 5

Question 1 — Glaciated Uplands

(a) Study OS map **Item A** of the Aviemore area.

Match the features of glaciated uplands shown below with the correct grid reference.

Features of glaciated uplands		
U-shaped valley	corrie	pyramidal peak

Choose from grid references			
954976	979976	957997	915980

3

(b) **Explain** the formation of a U-shaped valley.

You may use a diagram or diagrams in your answer.

4

Total marks   **7**

NOW ATTEMPT QUESTIONS 3, 4 AND 5

MARKS | DO NOT WRITE IN THIS MARGIN

DO NOT ATTEMPT THIS QUESTION IF YOU HAVE ALREADY ANSWERED QUESTION 1

Question 2 — Upland Limestone Areas

(a) Study OS map **Item B** of the Aber-craf area.

Match the following surface limestone features with the correct grid references.

Features of upland limestone areas		
pot hole	limestone pavement	intermittent drainage

Choose from grid references			
833164	817155	819163	814175

3

(b) **Explain** the formation of limestone pavement.

You may use a diagram or diagrams in your answer. 4

Total marks 7

NOW ATTEMPT QUESTIONS 3, 4 AND 5

MARKS | DO NOT WRITE IN THIS MARGIN

NOW ATTEMPT QUESTIONS 3, 4 AND 5

Question 3

Study OS map **Item A** of the Aviemore area.

Referring to map evidence, **explain** the ways in which the physical landscape has affected land use in the map extract area.

5

Question 4

Diagram Q4 — Synoptic Chart for 30 May 2013

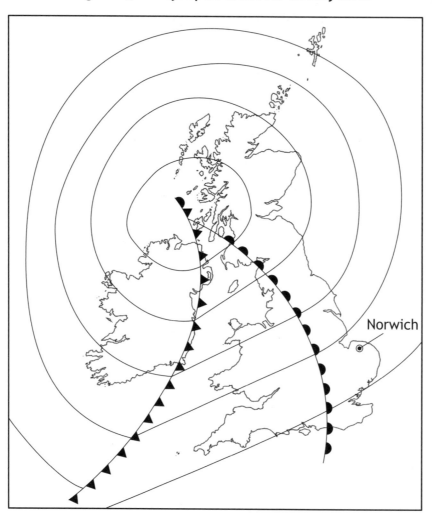

Study Diagram Q4 above.

Explain the changes that will take place in the weather at Norwich over the next twenty-four hours.

4

MARKS | DO NOT WRITE IN THIS MARGIN

Question 5

Diagram Q5A — Landscape Types   **Diagram Q5B — Land Uses**

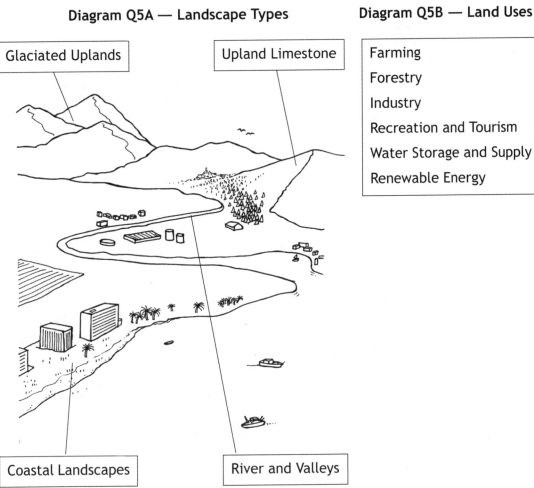

Glaciated Uplands

Upland Limestone

Coastal Landscapes

River and Valleys

Farming

Forestry

Industry

Recreation and Tourism

Water Storage and Supply

Renewable Energy

Look at Diagrams Q5A and Q5B.

Choose one landscape type you have studied from Diagram Q5A.

Select at least **two** land uses from Diagram Q5B and **explain** why these land uses are suitable for your chosen landscape type.

4

OS MAP ITEM C

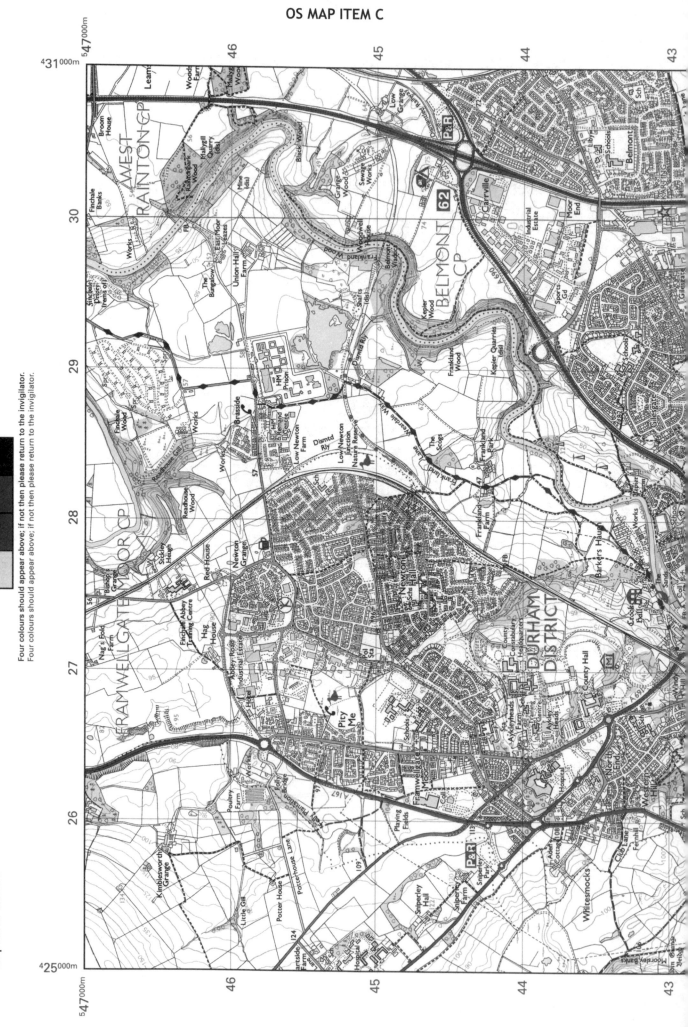

Extract No 1879/EXP308

1:25 000 Scale
Explorer Series

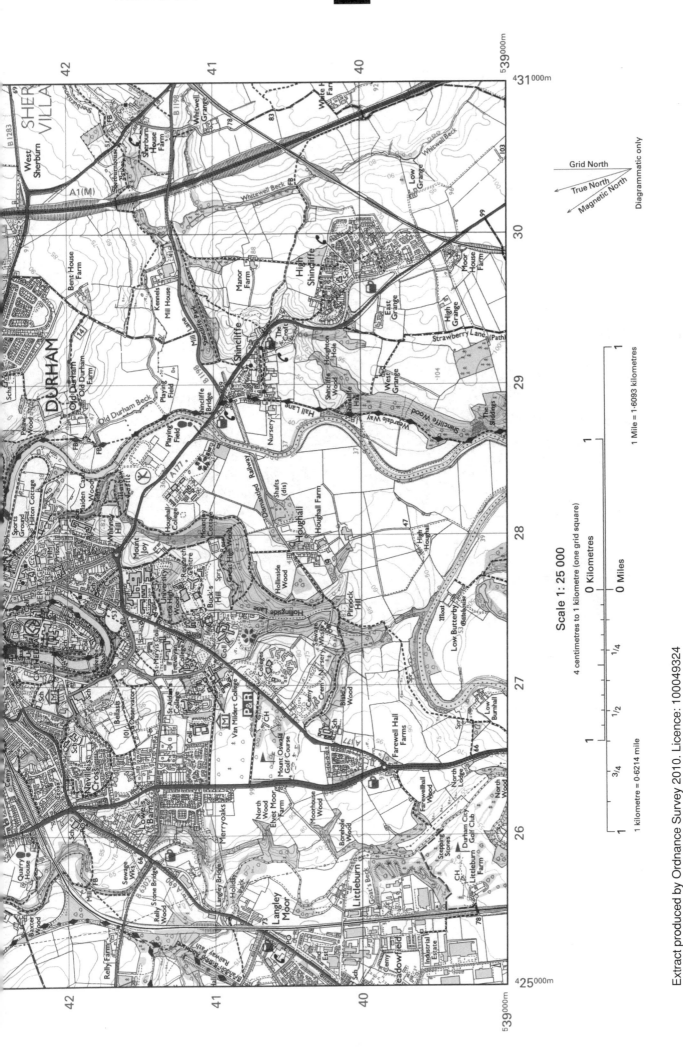

Grid North
True North
Magnetic North
Diagrammatic only

Scale 1: 25 000

4 centimetres to 1 kilometre (one grid square)

1   3/4   1/2   1/4   0 Kilometres   1   1

1   0 Miles   1

1 kilometre = 0·6214 mile   1 Mile = 1·6093 kilometres

MARKS | DO NOT WRITE IN THIS MARGIN

SECTION 2 — HUMAN ENVIRONMENTS — 20 marks
Attempt Questions 6, 7 and 8

Question 6

Study OS map **Item C** of the Durham area.

    (a) **Using Item C, give map evidence** to show that part of Durham's CBD is located in grid square 2742.

3

    (b) **Describe, in detail,** differences between the urban environments in grid squares 2642 and 2745.

5

Question 7

Diagram Q7 — Complexo do Alemão Shanty Town, Rio de Janeiro

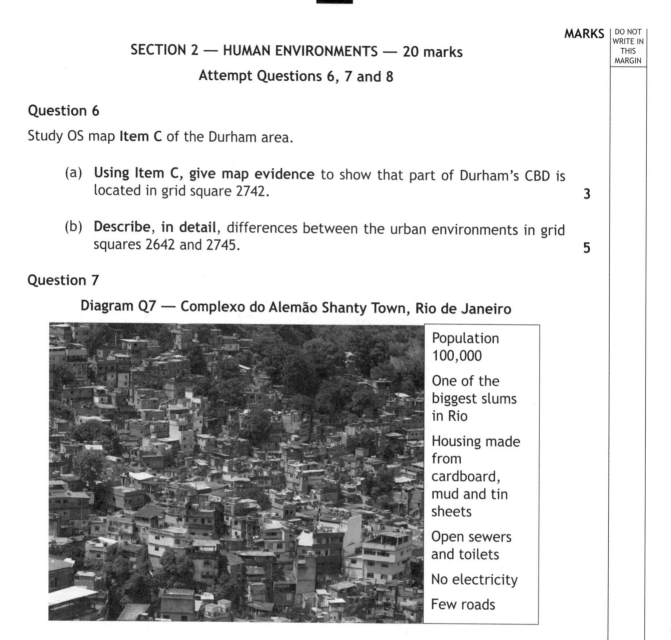

Population 100,000

One of the biggest slums in Rio

Housing made from cardboard, mud and tin sheets

Open sewers and toilets

No electricity

Few roads

Look at Diagram Q7.

For Rio de Janeiro, or a named developing world city you have studied, **describe** methods used by city authorities to improve living conditions in shanty towns.

6

MARKS | DO NOT WRITE IN THIS MARGIN

Question 8

Diagram Q8 — Modern Developments in Farming

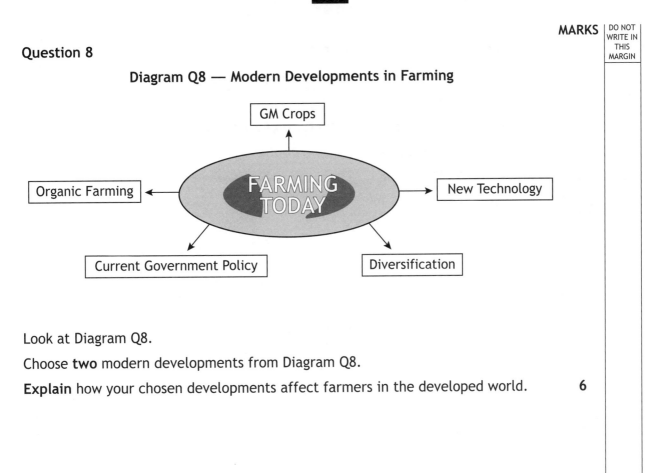

Look at Diagram Q8.

Choose **two** modern developments from Diagram Q8.

Explain how your chosen developments affect farmers in the developed world.   6

NOW GO TO SECTION 3

MARKS

DO NOT WRITE IN THIS MARGIN

SECTION 3 — GLOBAL ISSUES — 20 marks

Attempt any TWO questions

Question   9 — Climate Change

Question 10 — Impact of Human Activity on the Natural Environment

Question 11 — Environmental Hazards

Question 12 — Trade and Globalisation

Question 13 — Tourism

Question 14 — Health

MARKS

Question 9 — Climate Change

Diagram Q9 — Ice Melt 1979–2012 (months of June and July only)

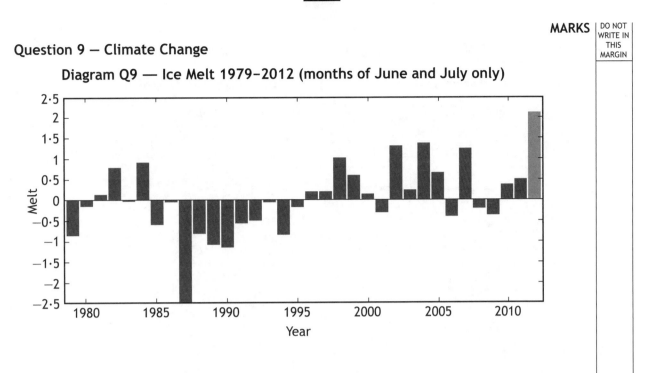

Study Diagram Q9.

(a) **Describe**, in detail, changes in ice melt between 1979 and 2012.   4

(b) Explain the effects of climate change on people and the environment.   6

Total marks 10

Question 10 — Impact of Human Activity on the Natural Environment

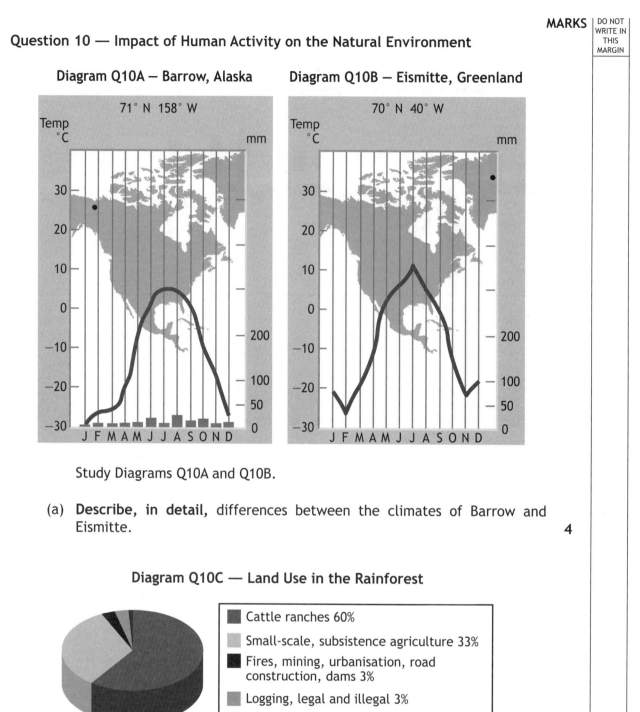

Diagram Q10A — Barrow, Alaska   Diagram Q10B — Eismitte, Greenland

Study Diagrams Q10A and Q10B.

(a) **Describe, in detail,** differences between the climates of Barrow and Eismitte.

4

Diagram Q10C — Land Use in the Rainforest

- Cattle ranches 60%
- Small-scale, subsistence agriculture 33%
- Fires, mining, urbanisation, road construction, dams 3%
- Logging, legal and illegal 3%
- Large-scale commercial agriculture including soybeans 1%

(b) Look at Diagram Q10C.

Explain how two of the land uses shown in Diagram Q10C can lead to degradation of the rainforest.

6

Total marks 10

MARKS   DO NOT WRITE IN THIS MARGIN

Question 11 — Environmental Hazards

Diagram Q11A — The World's Most Active Volcanoes

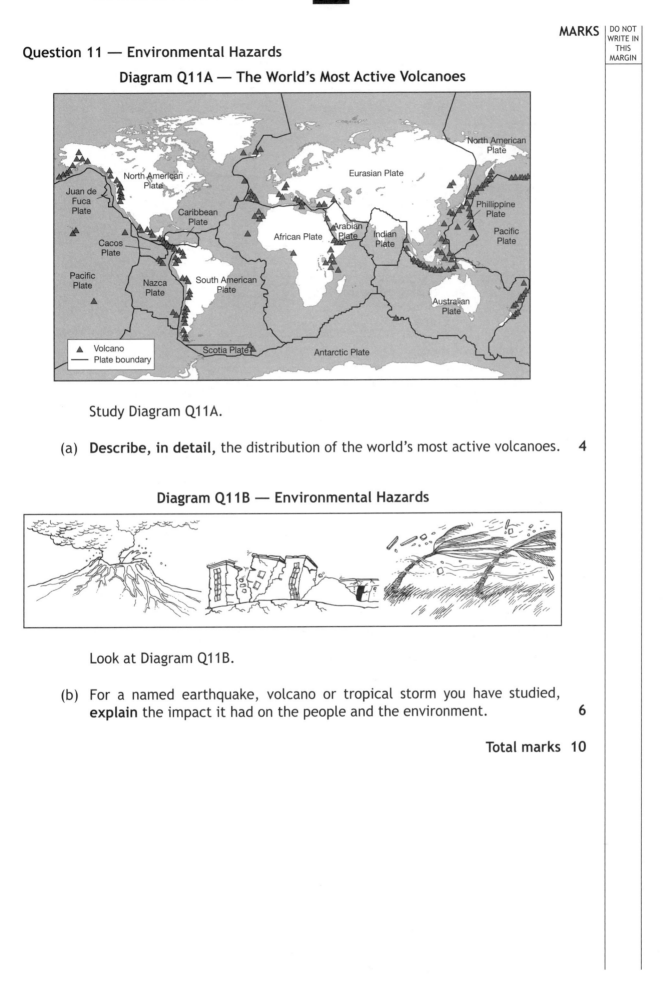

Study Diagram Q11A.

(a) **Describe, in detail,** the distribution of the world's most active volcanoes.   **4**

Diagram Q11B — Environmental Hazards

Look at Diagram Q11B.

(b) For a named earthquake, volcano or tropical storm you have studied, **explain** the impact it had on the people and the environment.   **6**

Total marks   **10**

MARKS | DO NOT WRITE IN THIS MARGIN

Question 12 — Trade and Globalisation

Diagram Q12A — Percentage Share of World Goods Production

1990	Developed countries	70%
	China	4%
	India	4%
2000	Developed countries	68%
	China	7%
	India	4%
2010	Developed countries	50%
	China	14%
	India	6%
2016 (projected)	Developed countries	45%
	China	18%
	India	7%

Study Diagram Q12A.

(a) **Describe, in detail,** the trends in percentage share of world goods production.   4

Diagram Q12B — Newspaper Headline

"The inequality in trade between rich and poor nations is now wider than it has ever been before"

Look at Diagram Q12B.

(b) **Explain** the causes of inequalities in trade between developed and developing countries.   6

Total marks 10

MARKS

Question 13 — Tourism

Diagram Q13 — Global Tourist Arrivals 1990–2011

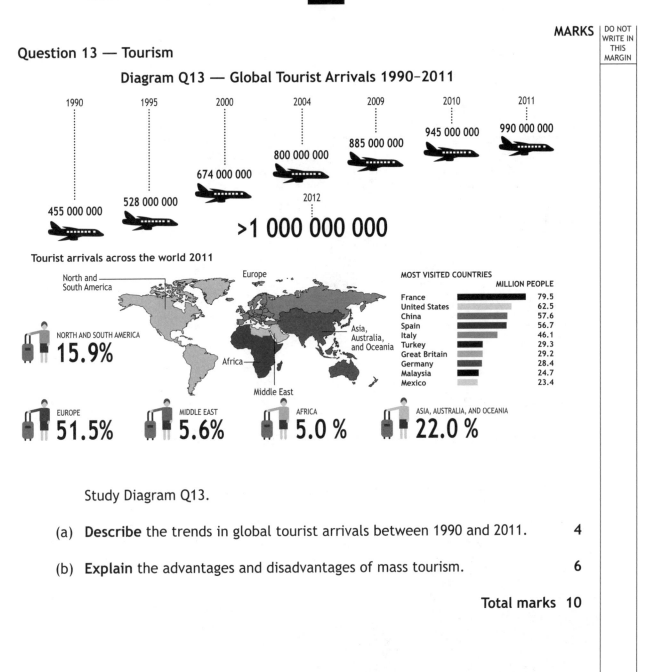

Study Diagram Q13.

(a) **Describe** the trends in global tourist arrivals between 1990 and 2011.   4

(b) **Explain** the advantages and disadvantages of mass tourism.   6

Total marks 10

MARKS | DO NOT WRITE IN THIS MARGIN

Question 14 — Health

Diagram Q14A — Adults Infected with HIV/AIDS

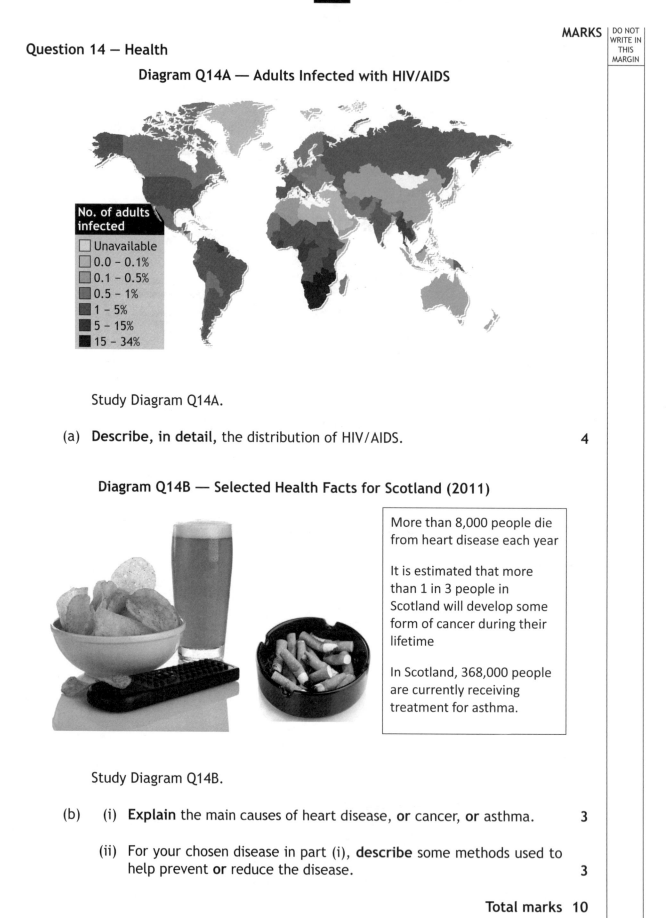

No. of adults infected
- ☐ Unavailable
- ☐ 0.0 – 0.1%
- ☐ 0.1 – 0.5%
- ☐ 0.5 – 1%
- ☐ 1 – 5%
- ☐ 5 – 15%
- ☐ 15 – 34%

Study Diagram Q14A.

(a) **Describe, in detail**, the distribution of HIV/AIDS.   4

Diagram Q14B — Selected Health Facts for Scotland (2011)

More than 8,000 people die from heart disease each year

It is estimated that more than 1 in 3 people in Scotland will develop some form of cancer during their lifetime

In Scotland, 368,000 people are currently receiving treatment for asthma.

Study Diagram Q14B.

(b) (i) **Explain** the main causes of heart disease, **or** cancer, **or** asthma.   3

   (ii) For your chosen disease in part (i), **describe** some methods used to help prevent **or** reduce the disease.   3

Total marks 10

[END OF QUESTION PAPER]

2013 Model Paper 2

N5

National
Qualifications
MODEL PAPER 2

Geography

Duration — 1 hour and 30 minutes

Total marks — 60

SECTION 1 — PHYSICAL ENVIRONMENTS — 20 marks

Attempt EITHER question 1 **or** question 2 AND questions 3, 4 and 5

SECTION 2 — HUMAN ENVIRONMENTS — 20 marks

Attempt questions 6, 7 and 8

SECTION 3 — GLOBAL ISSUES — 20 marks

Attempt any TWO of the following

Question   9 — Climate Change

Question 10 — Impact of Human Activity on the Natural Environment

Question 11 — Environmental Hazards

Question 12 — Trade and Globalisation

Question 13 — Tourism

Question 14 — Health

You should read the questions carefully.

Credit will always be given for appropriately labelled sketch maps and diagrams.

Use **blue** or **black** ink.

OS MAP ITEM A

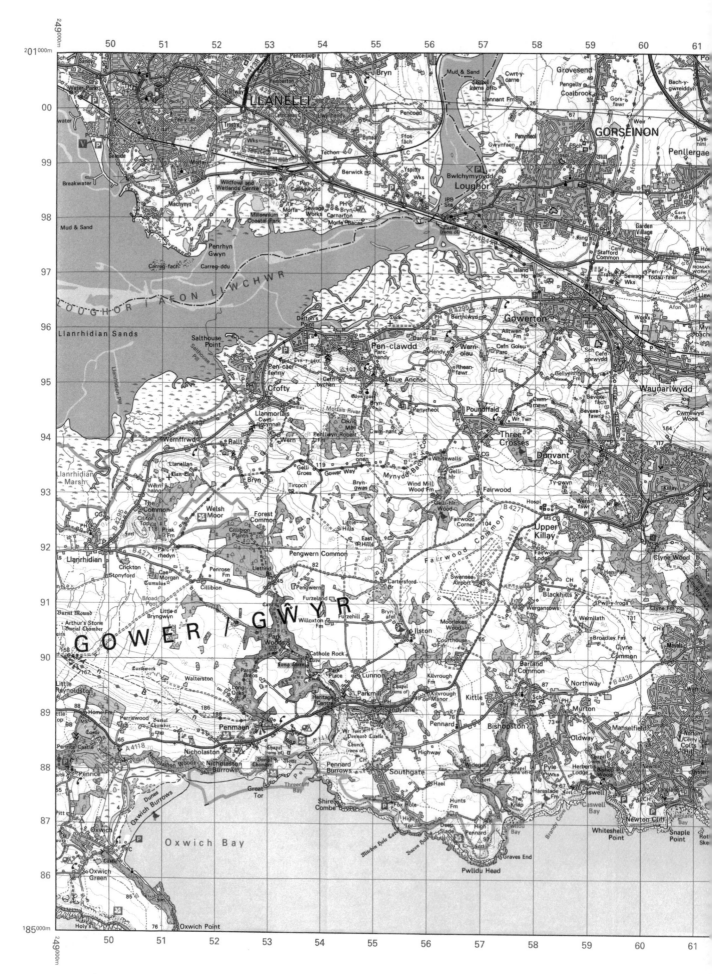

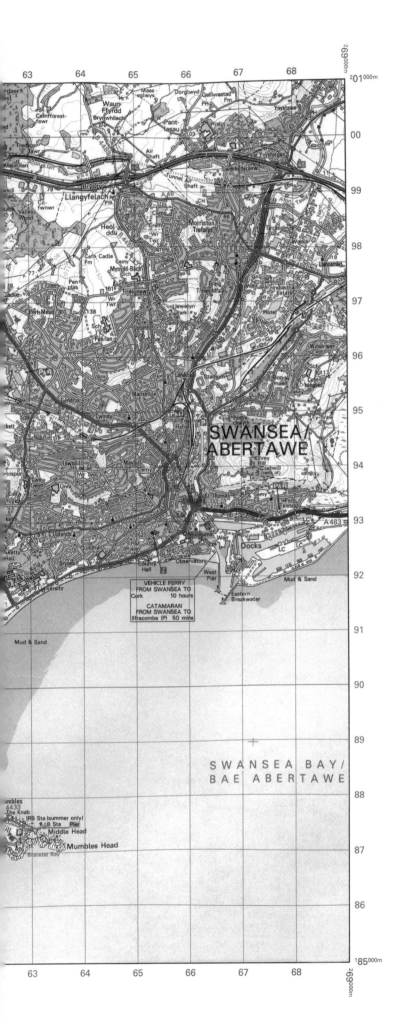

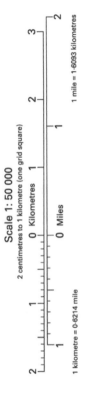

Scale 1: 50 000

2 centimetres to 1 kilometre (one grid square)

MARKS

DO NOT WRITE IN THIS MARGIN

SECTION 1 — PHYSICAL ENVIRONMENTS — 20 marks
Attempt EITHER Question 1 or Question 2
AND Questions 3, 4 and 5

Question 1 — Coastal Landscapes

(a) Study OS map **Item A** of the Swansea area.

Match the coastal features shown below with the correct grid reference.

Features of coastal landscapes		
bay	headland	stack

Choose from grid references			
636871	570863	592974	555869

3

(b) **Explain** the formation of caves, arches and stacks.

You may use a diagram or diagrams in your answer.

4

Total marks   **7**

NOW ATTEMPT QUESTIONS 3, 4 AND 5

MARKS | DO NOT WRITE IN THIS MARGIN

DO NOT ATTEMPT THIS QUESTION IF YOU HAVE ALREADY ANSWERED QUESTION 1

Question 2 — Rivers and Valleys

(a) Study OS map **Item A** of the Swansea area.

Match the river and valley features shown below with the correct grid reference.

Features of rivers and valleys		
confluence	meander	river flowing south

Choose from grid references			
670967	672973	584973	614965

3

(b) **Explain** the formation of a waterfall.

You may use a diagram or diagrams in your answer.

4

Total marks   7

NOW ATTEMPT QUESTIONS 3, 4 AND 5

MARKS

NOW ATTEMPT QUESTIONS 3, 4 AND 5

Question 3

Study OS map **Item A** of the Swansea area.

The area south of Swansea is a popular destination for tourists.

Using map evidence, explain the attractions of the physical landscape for tourists.   **5**

Question 4

Diagram Q4 — An Anticyclone over the UK

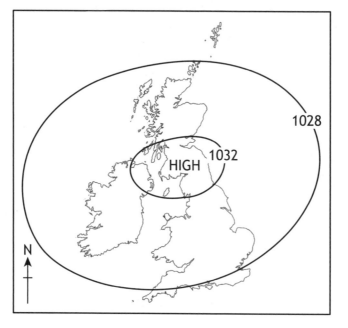

Look at Diagram Q4.

Describe similarities and differences in weather conditions caused by anticyclones in summer and winter.   **4**

MARKS | DO NOT WRITE IN THIS MARGIN

Question 5

Diagram Q5 — Land-Use Types and Land Uses

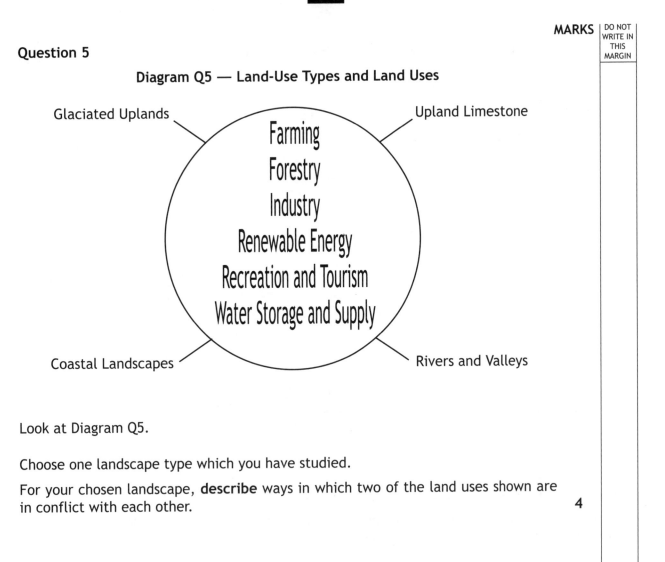

Glaciated Uplands

Upland Limestone

Farming
Forestry
Industry
Renewable Energy
Recreation and Tourism
Water Storage and Supply

Coastal Landscapes

Rivers and Valleys

Look at Diagram Q5.

Choose one landscape type which you have studied.

For your chosen landscape, **describe** ways in which two of the land uses shown are in conflict with each other.

4

SECTION 2 — HUMAN ENVIRONMENTS — 20 marks

Attempt Questions 6, 7 and 8

Question 6

Diagram Q6 — Selected Land-Use Zones in Swansea

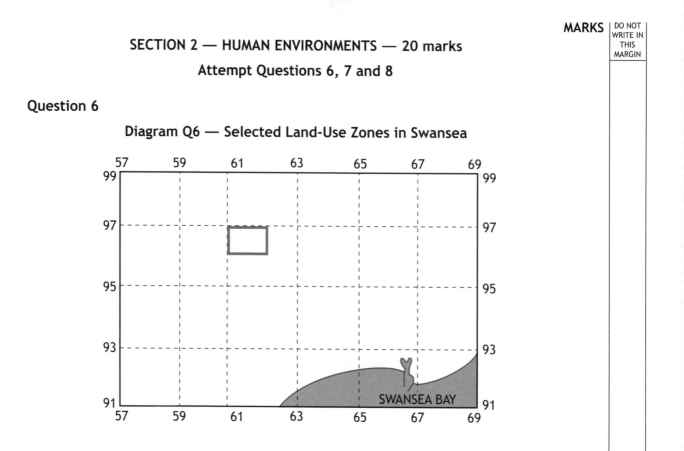

Study OS map **Item A** of the Swansea area and Diagram Q6.

It is proposed to build a shopping complex in square 6196. **Using map evidence, describe** the advantages and disadvantages of this site.

5

MARKS | DO NOT WRITE IN THIS MARGIN

Question 7

Diagram Q7 — World Population Distribution

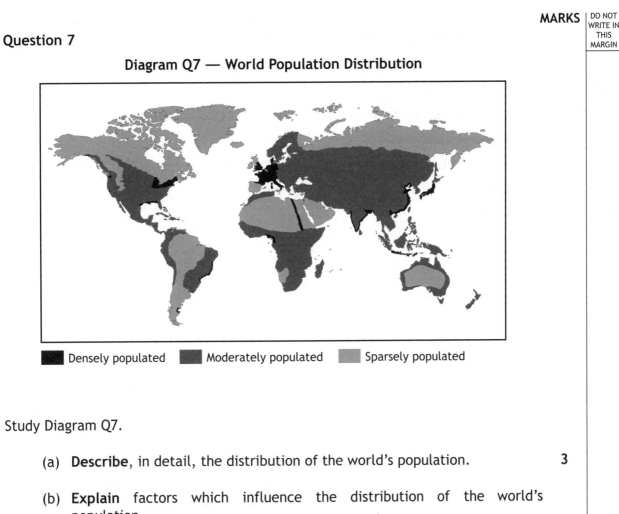

■ Densely populated   ■ Moderately populated   ■ Sparsely populated

Study Diagram Q7.

(a) **Describe**, in detail, the distribution of the world's population. **3**

(b) **Explain** factors which influence the distribution of the world's population.

You should refer to **both** human and physical factors in your answer. **6**

Total marks 9

Question 8

Diagram Q8 — Indicators of Development in the USA and Chad

Indicators	USA — Developed Country	Chad — Developing Country
Birth rate	13	40
Exports	steel, motor vehicles, telecommunications	oil, cattle, cotton
Life expectancy at birth	79	50
Adult literacy rate	99%	35%
GDP per capita (US $)	$49,000	$2000

Study Diagram Q8.

Choose **two** indicators from Diagram Q8.

Explain the reasons for the different levels of development in the USA and Chad. **6**

NOW GO TO SECTION 3

SECTION 3 — GLOBAL ISSUES — 20 marks

Attempt any TWO questions

Question　9 — Climate Change

Question 10 — Impact of Human Activity on the Natural Environment

Question 11 — Environmental Hazards

Question 12 — Trade and Globalisation

Question 13 — Tourism

Question 14 — Health

MARKS | DO NOT WRITE IN THIS MARGIN

Question 9 — Climate Change

Diagram Q9 — Average Global Temperatures 1880–2020

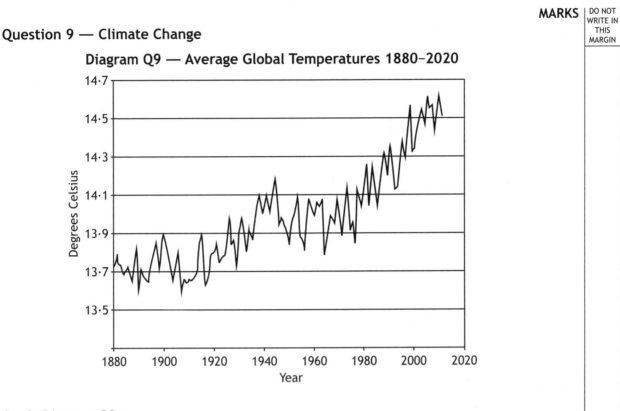

Study Diagram Q9.

(a) **Describe, in detail,** the changes in average global temperatures between 1880 and 2020 (predicted).   **4**

(b) **Explain** the physical and human causes of climate change.   **6**

Total marks 10

MARKS | DO NOT WRITE IN THIS MARGIN

Question 10 — Impact of Human Activity on the Natural Environment

Diagram Q10 — Amazon Deforestation per Year, 1988–2010

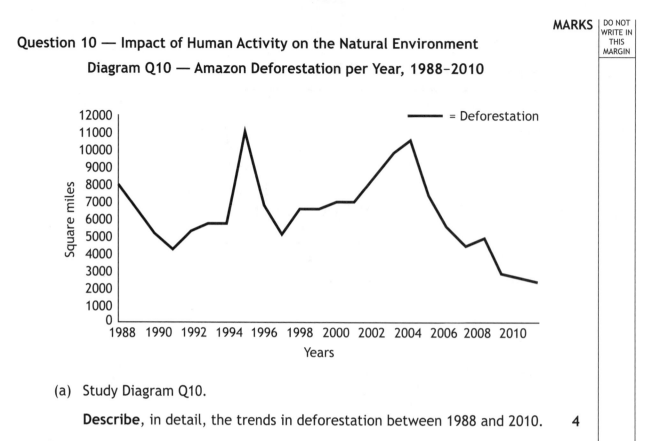

(a) Study Diagram Q10.

Describe, in detail, the trends in deforestation between 1988 and 2010.   4

(b) **Explain** ways in which the destruction of the rainforest can be reduced.   6

Total marks 10

MARKS | DO NOT WRITE IN THIS MARGIN

Question 11 — Environmental Hazards

Diagram Q11 — Distribution of Tropical Storms

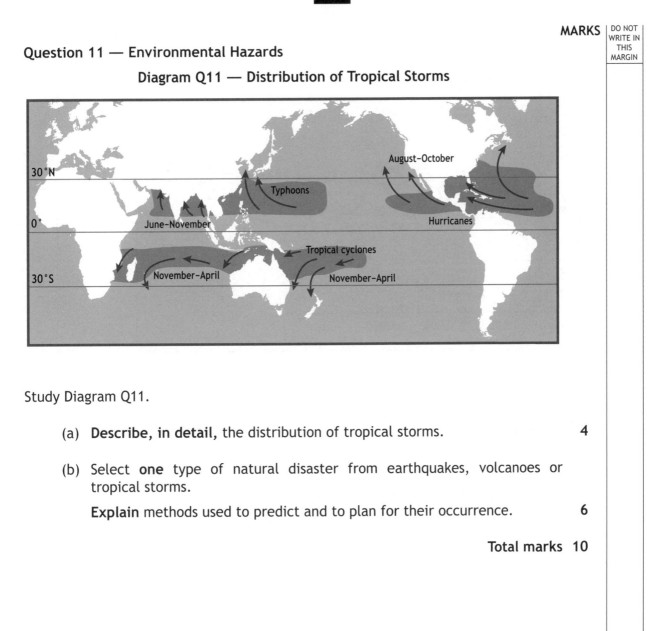

Study Diagram Q11.

(a) **Describe, in detail,** the distribution of tropical storms.   4

(b) Select **one** type of natural disaster from earthquakes, volcanoes or tropical storms.

Explain methods used to predict and to plan for their occurrence.   6

Total marks 10

MARKS | DO NOT WRITE IN THIS MARGIN

Question 12 — Trade and Globalisation

Diagram Q12A — Share of World Exports

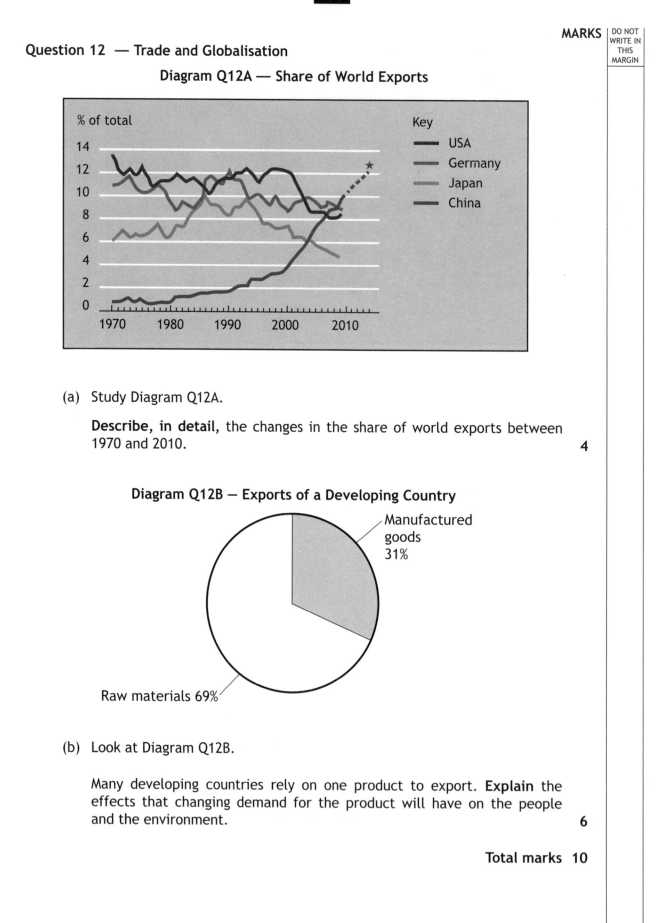

(a) Study Diagram Q12A.

Describe, in detail, the changes in the share of world exports between 1970 and 2010.

4

Diagram Q12B — Exports of a Developing Country

Manufactured goods 31%

Raw materials 69%

(b) Look at Diagram Q12B.

Many developing countries rely on one product to export. **Explain** the effects that changing demand for the product will have on the people and the environment.

6

Total marks 10

MARKS | DO NOT WRITE IN THIS MARGIN

Question 13 — Tourism

Diagram Q13 — Facts on Selected National Parks in the UK

National Park name	Year of designation	Population	Visitors per year (million)	Visitor days per year (million)	Visitor spend per year (million)
Brecon Beacons	1957	32,000	4·15	5	£197
Cairngorms	2003	17,000	1·5	3·1	£185
Lake District	1951	42,200	15·8	23·1	£952
Loch Lomond and the Trossachs	2002	15,600	4	7	£190
Pembrokeshire coast	1952	22,600	4·2	13	£498
Snowdonia	1951	25,482	4·27	10·4	£396

Study Diagram Q13.

(a) **Describe, in detail,** the differences between the selected National Parks. 4

(b) For a named area you have studied, **explain** methods used to manage problems caused by mass tourism. 6

Total marks 10

MARKS | DO NOT WRITE IN THIS MARGIN

Question 14 — Health

Diagram Q14 — Cholera outbreaks 2010–2011

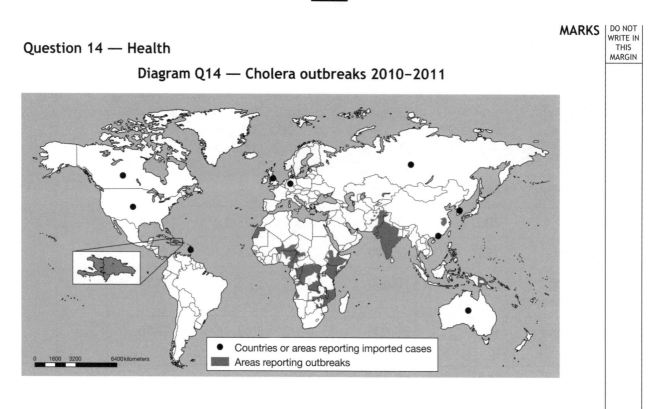

Study Diagram Q14.

(a) **Describe, in detail,** the distribution of cholera outbreaks in 2010 and 2011.   **4**

(b) For cholera, malaria, kwashiorkor **or** pneumonia, **explain** the main causes of your chosen disease and **describe** some methods used to help prevent or reduce the disease.   **6**

Total marks 10

[END OF QUESTION PAPER]

NATIONAL 5

2013 Model Paper 3

National
Qualifications
MODEL PAPER 3

Geography

Duration — 1 hour and 30 minutes

Total marks — 60

SECTION 1 — PHYSICAL ENVIRONMENTS — 20 marks

Attempt EITHER question 1 **or** question 2 AND questions 3, 4 and 5

SECTION 2 — HUMAN ENVIRONMENTS — 20 marks

Attempt questions 6, 7 and 8

SECTION 3 — GLOBAL ISSUES — 20 marks

Attempt any TWO of the following

Question 9 — Climate Change

Question 10 — Impact of Human Activity on the Natural Environment

Question 11 — Environmental Hazards

Question 12 — Trade and Globalisation

Question 13 — Tourism

Question 14 — Health

You should read the questions carefully.

Credit will always be given for appropriately labelled sketch maps and diagrams.

Use **blue** or **black** ink.

OS MAP ITEM A

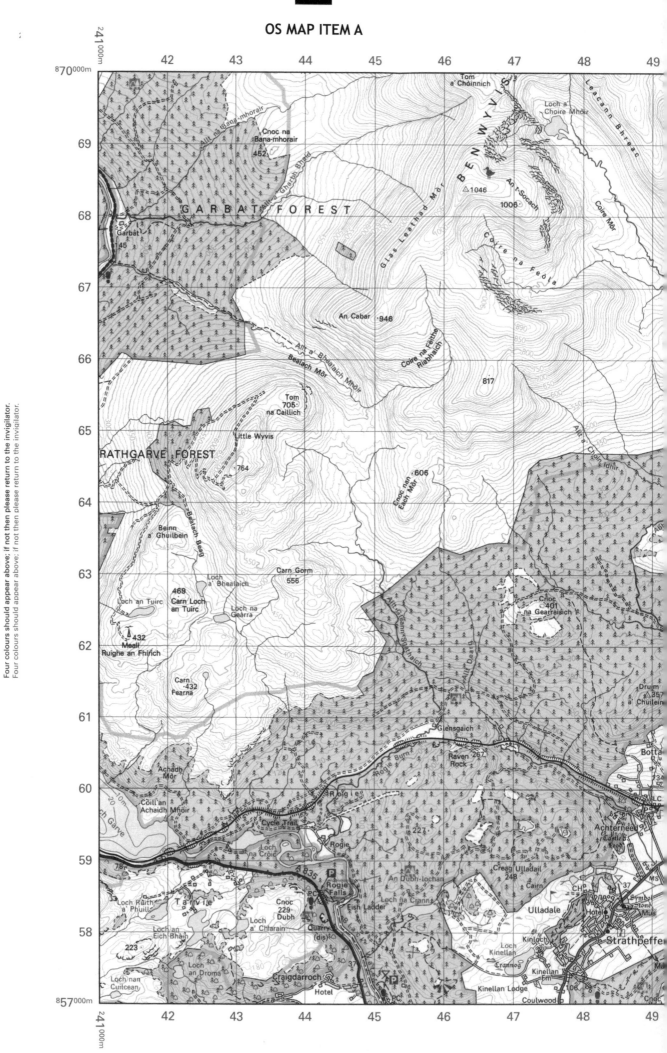

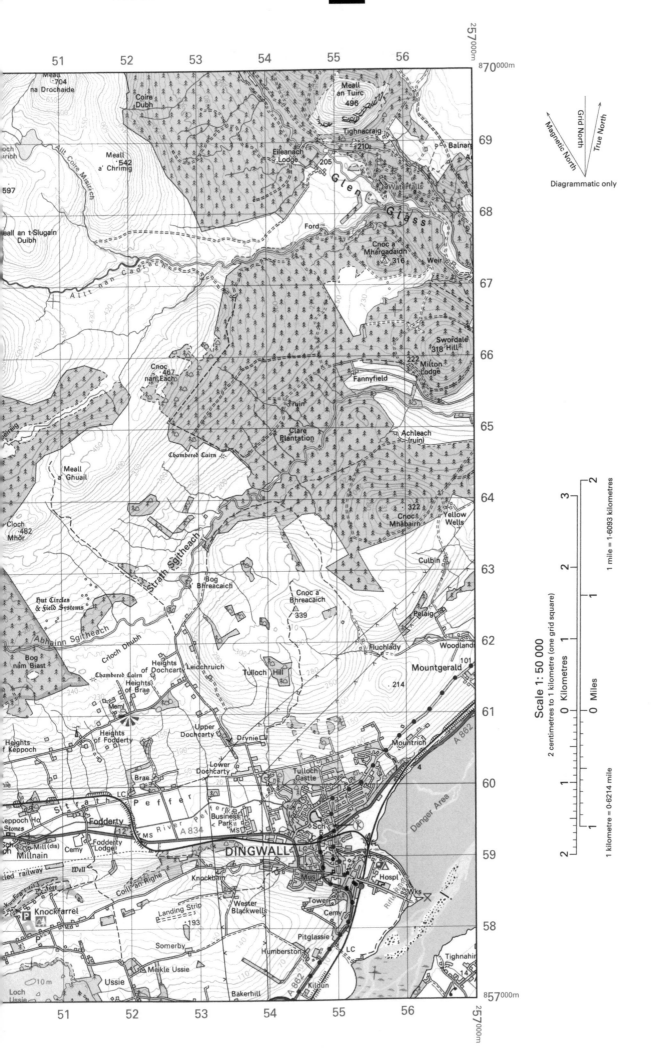

Scale 1: 50 000

2 centimetres to 1 kilometre (one grid square)

MARKS | DO NOT WRITE IN THIS MARGIN

SECTION 1 — PHYSICAL ENVIRONMENTS — 20 marks

Attempt EITHER Question 1 or Question 2
AND Questions 3, 4 and 5

Question 1 — Glaciated Uplands

Study OS map **Item A** of the Dingwall area.

(a) **Match** the glaciated uplands features shown below with the correct grid reference.

Features of glaciated uplands		
arête	corrie	U-shaped valley

Choose from grid references			
467677	472693	435663	525594

3

(b) **Explain** the formation of a corrie.

You may use a diagram or diagrams in your answer.   **4**

Total marks   **7**

NOW ATTEMPT QUESTIONS 3, 4 AND 5

MARKS

DO NOT ATTEMPT THIS QUESTION IF YOU HAVE ALREADY ANSWERED QUESTION 1

Question 2 — Rivers and Valleys

Study OS map **Item A** of the Dingwall area.

(a) **Match** the following river and valley features to the correct grid references.

Features of rivers and valleys		
V-shaped valley	meander	tributary

Choose from grid references			
528595	442591	447585	473657

3

(b) **Explain** the formation of a V-shaped valley.

You may use a diagram or diagrams in your answer.

4

Total marks 7

NOW ATTEMPT QUESTIONS 3, 4 AND 5

MARKS | DO NOT WRITE IN THIS MARGIN

NOW ATTEMPT QUESTIONS 3, 4 AND 5

Question 3

Diagram Q3A — Cross-section from GR 466658 to GR 510580

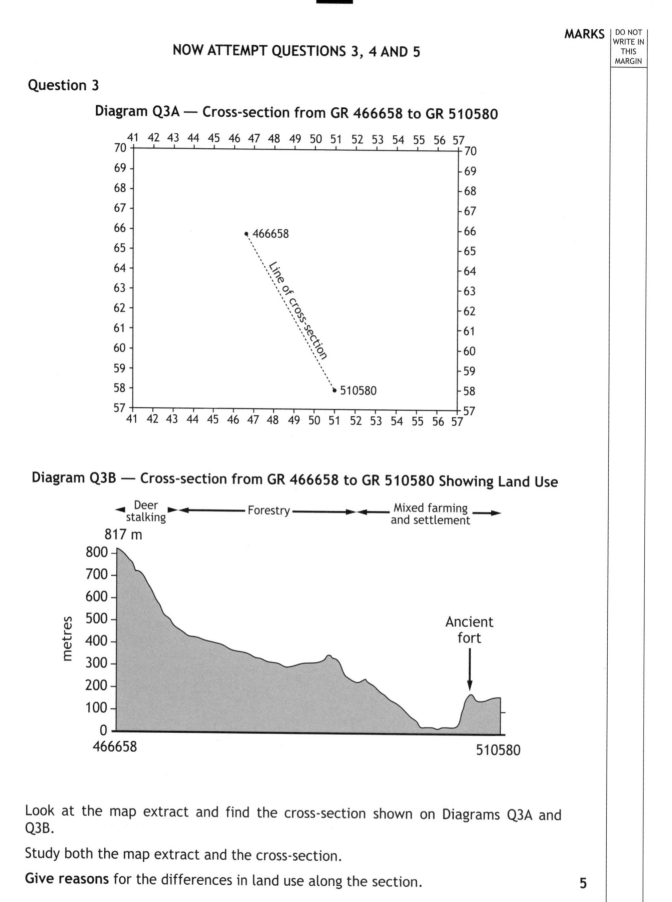

Diagram Q3B — Cross-section from GR 466658 to GR 510580 Showing Land Use

Look at the map extract and find the cross-section shown on Diagrams Q3A and Q3B.

Study both the map extract and the cross-section.

Give reasons for the differences in land use along the section.

5

Question 4

Diagram Q4 — Synoptic Chart, 26 February 2013

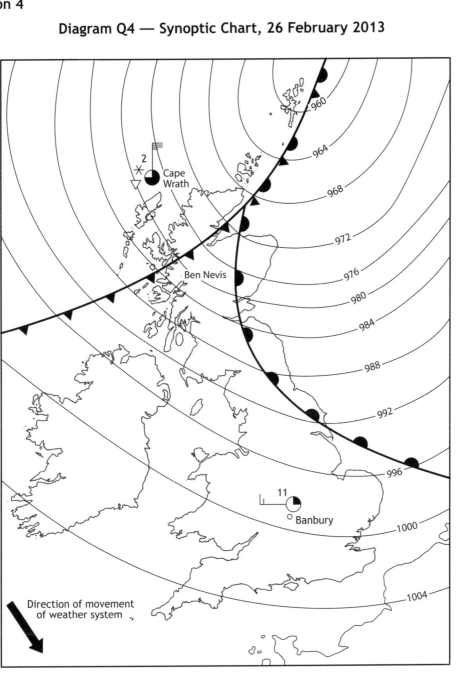

Study Diagram Q4.

Use the information in Diagram Q4 to **give reasons** for the differences in the weather conditions between Banbury and Cape Wrath.

4

MARKS | DO NOT WRITE IN THIS MARGIN

Question 5

Diagram Q5 — Landscape Types and Selected Land Uses

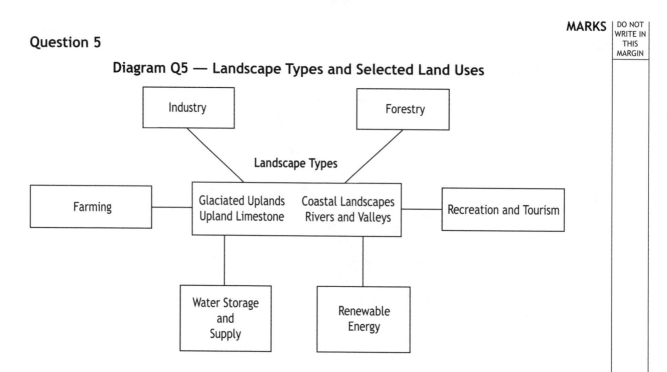

Look at Diagram Q5.

Choose one landscape type which you have studied.

For your chosen landscape, **give reasons** why one of the land uses from Diagram Q5 is found there.

4

MARKS | DO NOT WRITE IN THIS MARGIN

SECTION 2 — HUMAN ENVIRONMENTS — 20 marks
Attempt Questions 6, 7 and 8

Question 6

Study OS map **Item A** of the Dingwall area.

There is a plan to build a new housing estate in grid square 5558.

Using map evidence, give the advantages and disadvantages of this proposal.   **5**

Question 7

Diagram Q7 — GDP per Capita and Number of Births per Woman in Selected Countries

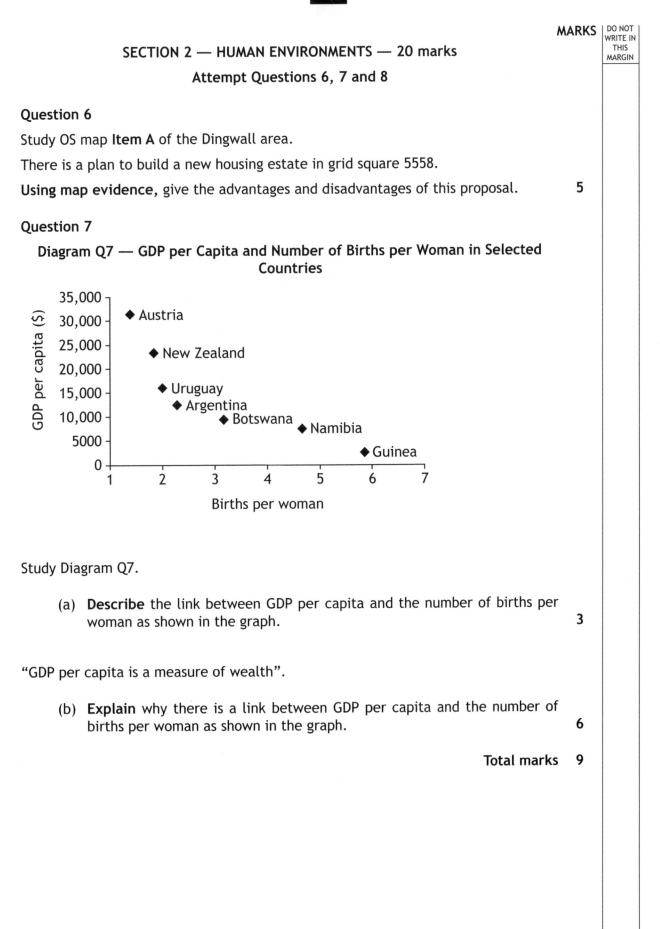

Study Diagram Q7.

(a) **Describe** the link between GDP per capita and the number of births per woman as shown in the graph.   **3**

"GDP per capita is a measure of wealth".

(b) **Explain** why there is a link between GDP per capita and the number of births per woman as shown in the graph.   **6**

Total marks   9

Question 8

Diagram Q8 — World Population Growth (projected)

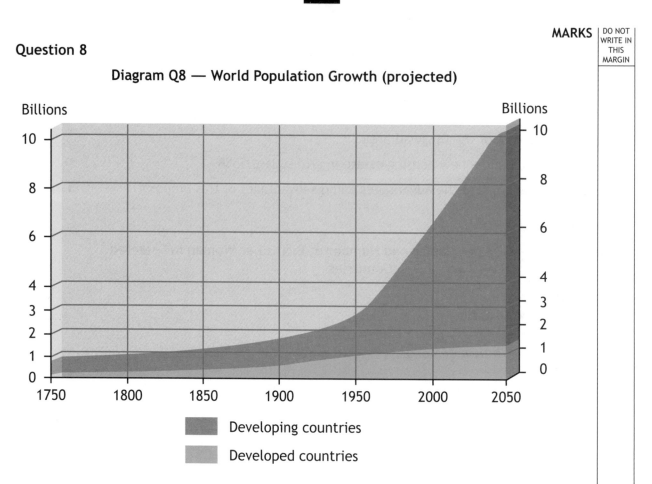

Look at Diagram Q8.

What are the implications of the population growth shown on Diagram Q8 for developing and developed countries?

6

NOW GO TO SECTION 3

MARKS | DO NOT WRITE IN THIS MARGIN

SECTION 3 — GLOBAL ISSUES — 20 marks

Attempt any TWO questions

Question 9 — Climate Change

Question 10 — Impact of Human Activity on the Natural Environment

Question 11 — Environmental Hazards

Question 12 — Trade and Globalisation

Question 13 — Tourism

Question 14 — Health

Question 9 — Climate Change

Diagram Q9A — Changes in Carbon Dioxide and Temperatures, 1900–2010

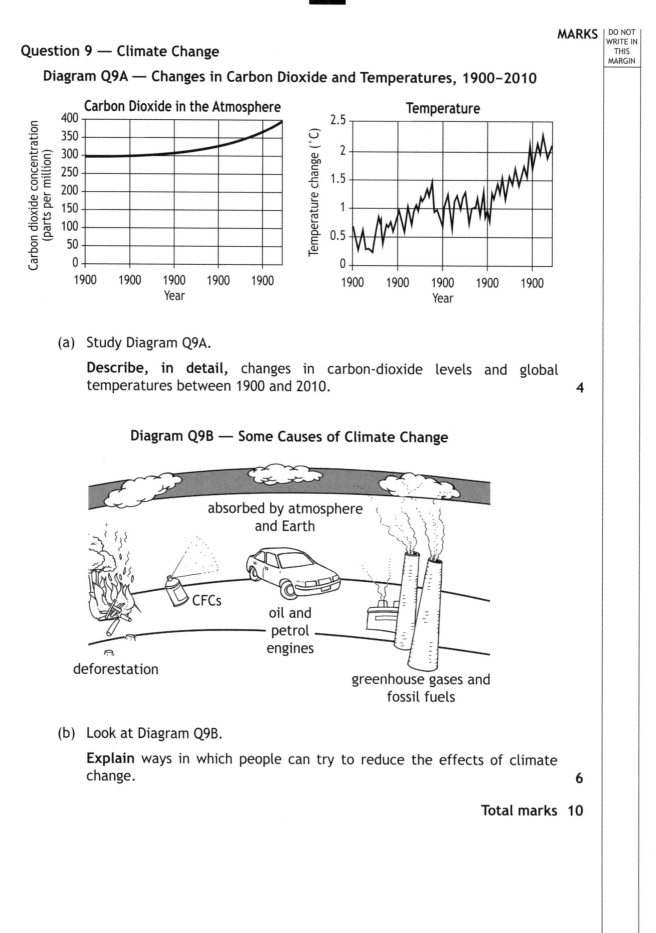

(a) Study Diagram Q9A.

Describe, in detail, changes in carbon-dioxide levels and global temperatures between 1900 and 2010.

4

Diagram Q9B — Some Causes of Climate Change

(b) Look at Diagram Q9B.

Explain ways in which people can try to reduce the effects of climate change.

6

Total marks   10

Question 10 — Impact of Human Activity on the Natural Environment

Diagram Q10 — SE Asia: Changes in Forest Cover, 1985–2010

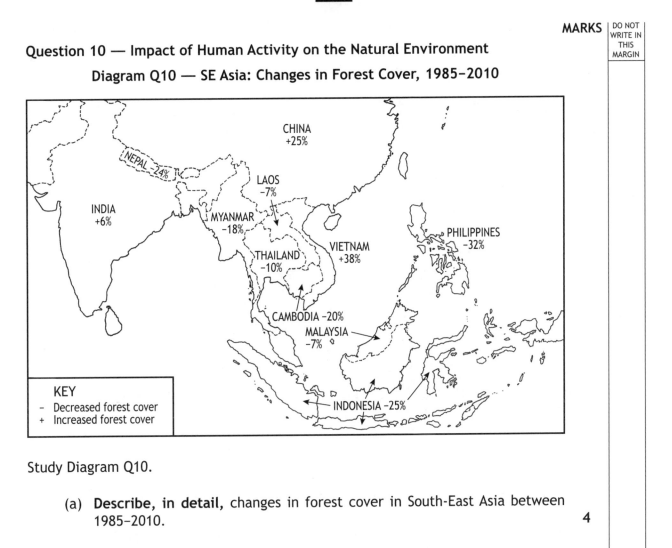

Study Diagram Q10.

(a) **Describe, in detail,** changes in forest cover in South-East Asia between 1985–2010.

4

(b) **Explain** the advantages and disadvantages of deforestation in equatorial areas.

In your answer, you should refer to an area you have studied.

6

Total marks 10

Question 11 — Environmental Hazards

Diagram Q11 — Distribution of Earthquakes, October 2012

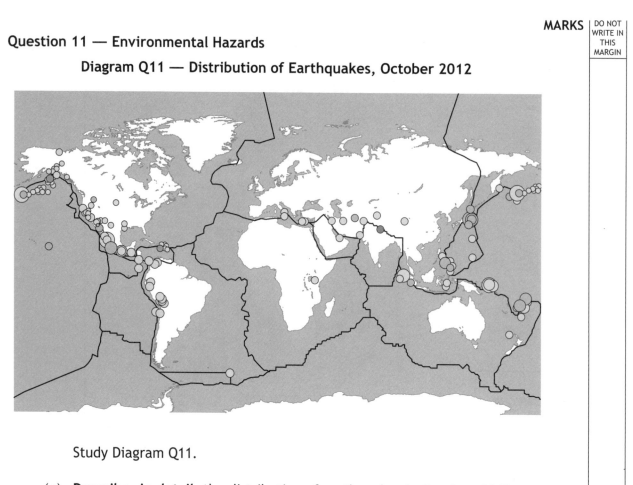

Study Diagram Q11.

(a) **Describe, in detail,** the distribution of earthquakes in October, 2012.   **4**

(b) For a named environmental hazard you have studied, **explain** the ways in which aid can lessen the impact of environmental hazards.   **6**

Total marks 10

MARKS | DO NOT WRITE IN THIS MARGIN

Question 12 — Trade and Globalisation

Diagram Q12A — A Selected Country's Imports and Exports

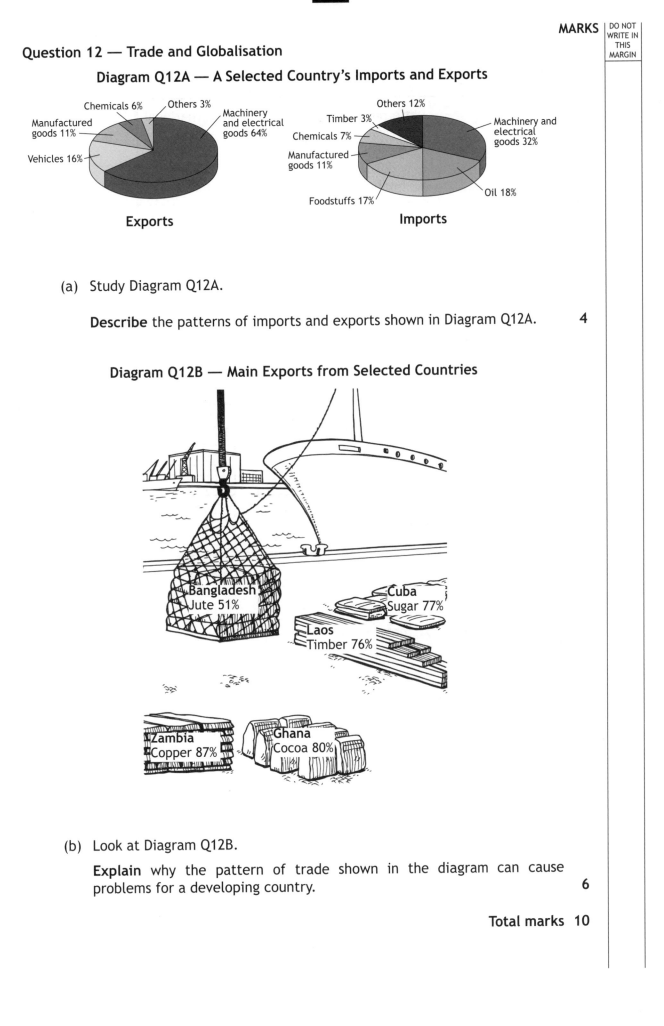

Exports

Imports

(a) Study Diagram Q12A.

Describe the patterns of imports and exports shown in Diagram Q12A.   **4**

Diagram Q12B — Main Exports from Selected Countries

Bangladesh Jute 51%

Cuba Sugar 77%

Laos Timber 76%

Zambia Copper 87%

Ghana Cocoa 80%

(b) Look at Diagram Q12B.

Explain why the pattern of trade shown in the diagram can cause problems for a developing country.   **6**

Total marks 10

MARKS DO NOT WRITE IN THIS MARGIN

Question 13 — Tourism

Diagram Q13 — International Tourist Arrivals, 2008–2010

World position	Country	International arrivals of tourists in 2010 (millions)	International arrivals of tourists in 2009 (millions)	International arrivals of tourists in 2008 (millions)
1	France	76·3	76·8	79·2
2	Spain	60·1	55·2	57·9
3	USA	59·7	55·1	57·2
4	China	55·7	50·9	53·0
5	Italy	53·6	43·2	42·7
6	United Kingdom	28·1	28·2	30·1
7	Turkey	27·0	25·5	25·0
8	Germany	26·9	24·2	24·9
9	Malaysia	24·6	23·6	22·1
10	Mexico	22·4	21·5	22·6
Worldwide total		940	882	917

Study Diagram Q13.

(a) **Describe, in detail,** patterns in international tourist arrivals between 2008 and 2010. 4

(b) **Explain** ways in which eco-tourism helps the people and the environment of a developing country.

In your answer, you should refer to a country you have studied. 6

Total marks 10

MARKS | DO NOT WRITE IN THIS MARGIN

Question 14 — Health

Diagram Q14 — Malaria Cases in Selected South African Provinces

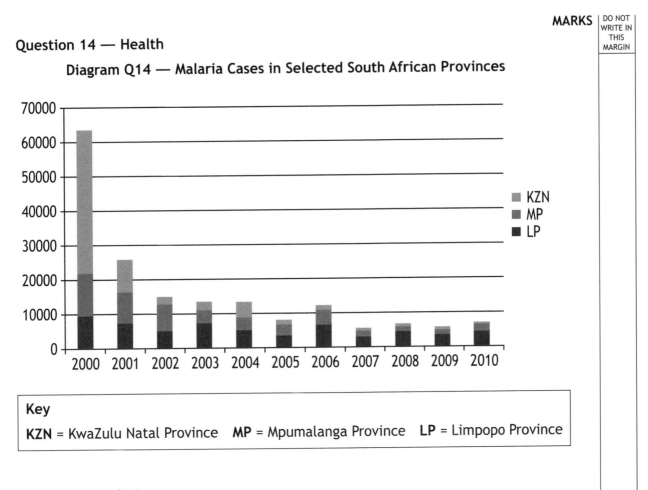

Key

KZN = KwaZulu Natal Province   **MP** = Mpumalanga Province   **LP** = Limpopo Province

Study Diagram Q14.

(a) **Describe, in detail,** the changes in malaria cases in the selected provinces of South Africa.

4

(b) For malaria, kwashiorkor, pneumonia or cholera, **explain** methods used by countries to control the spread of the disease.

6

Total marks 10

[END OF QUESTION PAPER]

SQA AND HODDER GIBSON NATIONAL 5 GEOGRAPHY 2013

General Marking Principles for National 5 Geography

Questions that ask candidates to *Describe* . . . (4–6 marks)

Candidates must make a number of relevant, factual points. These should be key points. The points do not need to be in any particular order. Candidates may provide a number of straightforward points or a smaller number of developed points, or a combination of these.

Up to the total mark allocation for this question:

- **One mark** should be given for each accurate relevant point.
- **Further marks** should be given for development and exemplification.

Question: Describe, in detail, the effects of two of the factors shown. (Modern factors affecting farming).

Example:

New technology has led to increased crop yields *(one mark)*, leading to better profits for some farmers *(a second mark for development)*.

Questions that ask candidates to *Explain* . . . (4–6 marks)

Candidates must make a number of points that make the process/situation plain or clear, for example by showing connections between factors or causal relationships between events or processes. These should be key reasons and may include theoretical ideas. There is no need for any prioritising of these reasons. Candidates may provide a number of straightforward reasons or a smaller number of developed reasons, or a combination of these. The use of the command word 'explain' will generally be used when candidates are required to demonstrate knowledge and understanding. However, depending on the context of the question the command words 'give reasons' may be substituted.

If candidates produce fully labelled diagrams they may be awarded up to full marks if the diagrams are sufficiently accurate and detailed.

Up to the total mark allocation for this question:

- **One mark** should be given for each accurate relevant point.
- **Further marks** should be given for developed explanations.

Question: Explain the formation of a U-shaped valley.

Example:
A glacier moves down a main valley which it erodes *(1 mark)* by plucking, where the ice freezes on to fragments of rock and pulls them away. *(second mark for development)*.

Questions that ask candidates to *Give reasons* . . . (4–6 marks)

Candidates must make a number of points that make the process/situation plain or clear, for example by showing connections between factors or causal relationships between events or processes. These should be key reasons and may include theoretical ideas. There is no need for any prioritising of these reasons. Candidates may provide a number of straightforward reasons or a smaller number of developed reasons, or a combination of these. The use of

the command words 'give reasons' will generally be used when candidates are required to use information from sources. However, depending on the context of the question the command word 'explain' may be substituted.

Up to the total mark allocation for this question:

- **One mark** should be given for each accurate relevant point.
- **Further marks** should be given for developed reasons.

Question: Give reasons for the differences in the weather conditions between Belfast and Stockholm.

Example:
In Stockholm it is dry but in Belfast it is wet because Stockholm is in a ridge of high pressure whereas Belfast is in a depression *(one mark)*. Belfast is close to the warm front and therefore experiencing rain *(second mark for development)*.

Questions that ask candidates to *Match* (3–4 marks)

Candidates must match two sets of variables by using their map interpretation skills.

Up to the total mark allocation for this question:

One mark should be given for each correct answer.

Question: Match the letters A to C with the correct features.

Example: A = Forestry *(1 mark)*

Questions that ask candidates to *Give map evidence* (3–4 marks)

Candidates must look for evidence on the map and make clear statements to support their answer.

Up to the total mark allocation for this question:

Question: Give map evidence to show that part of Coventry's CBD is located in grid square 3379.

Example: Many roads meet in this square *(1 mark)*.

Questions that ask candidates to *Give advantages and/or disadvantages* (4–6 marks)

Candidates must select relevant advantages or disadvantages of a proposed development and show their understanding of their significance to the proposal. Answers may give briefly explained points or a smaller number of points which are developed to warrant further marks.

Up to the total mark allocation for this question:

- **One mark** should be given for each accurate relevant point.
- **Further marks** should be given for developed points.
- Marks should be awarded for accurate map evidence.

Question: Give either advantages or disadvantages of this location for a shopping centre. You must use map evidence to support your answer.

Example: There are roads and motorways close by allowing the easy delivery of goods *(1 mark)* and access for customers *(1 mark for development)*, eg the A46, M6 and M69.

2013 Specimen Question Paper: Marking Instructions for each question

Section 1: Physical Environments

Section 1		General Marking Instructions for this type of question	Max mark	Specific Marking Instructions for this question
1.	(a)		3	U-shaped valley 483685 Corrie 467677 Truncated spur 476683
	(b)	A well annotated diagram could obtain full marks.	4	eg A glacier moves down a main valley which it erodes by plucking, where the ice freezes on to fragments of rock and pulls them away and abrasion, where rock fragments embedded in the ice scrape the land surface. As a result the valley becomes deeper, straighter and wider.
2.	(a)		3	V-shaped valley 457668 Meander 522623 River flowing NW 435663
	(b)	A well annotated diagram could obtain full marks.	4	eg At a river meander, water is pushed towards the outside of the bend causing erosion. The slower flow of water on the inside bend causes deposition. Over time erosion narrows the neck of the meander. In time, usually during a flood, the river will cut right through the neck. The fastest current is now in the centre of the river and deposition occurs next to the banks eventually blocking off the meander to leave an ox-bow lake.
3.			3	A = Forestry B = River Peffer C = Dismantled railway
4.		Candidates must make a number of points that make the process/ situation plain or clear, for example by showing connections between factors or causal relationships between events or processes. These should be key reasons and may include theoretical ideas. Candidates may provide a number of straightforward reasons or a smaller number of developed reasons, or a combination of these.	4	Possible answers might include: Wind direction at Stockholm is NW whereas at Belfast it is South due to the different alignment of the isobars, winds circulate in an anticlockwise direction around a depression but clockwise around an area of high pressure. Belfast has stronger winds because the isobars are closer together. In Stockholm it is dry but in Belfast it is wet because Stockholm is in a ridge of high pressure whereas Belfast is in a depression. Belfast is close to the warm front and therefore experiencing rain. There are 8 oktas of cloud cover in Belfast because it is close to the warm front, whereas Stockholm is not yet affected by the clouds associated with the advancing warm front. Temperatures in Belfast are warmer than in Stockholm as it in the warm sector. Clear skies in a high pressure in January will lead to cold temperatures in Stockholm.
5.		Answers will depend on the chosen landscape type.	6	At least <u>two</u> land uses must be described for full marks. Possible answers for Upland Limestone might include: Farming activities are often in conflict with walkers' rights of access, walkers leaving gates open, dogs chasing sheep, stone walls damaged. Careless tourists might drop litter. Quarrying is often in conflict with tourism because it spoils the appearance of the countryside, for example Swinden Quarry in the Yorkshire Dales; quarrying also leads to heavy traffic on narrow roads, creates noise and air pollution, all of which might put tourists off visiting the area.

Section 2: Human Environments

Section 2		General Marking Instructions for this type of question	Max mark	Specific Marking Instructions for this question
6.	(a)		3	Many roads meet in this square, there is a tourist information centre, many churches, a cathedral, bus station.
	(b)	Full marks may be awarded for either advantages or disadvantages or a combination of both.	5	**Advantages:** The land is flat so easy to build on there is room for expansion there are roads and motorways close by allowing the easy delivery of goods and access for customers eg the A46, M6 and M69 the land is cheaper on the outskirts the housing close by can supply a workforce eg Walsgrave on Sowe the city of Coventry can supply a large amount of customers. **Disadvantages:** Roads take up a large amount of land so only a limited amount of land available to build on and the road junctions could become congested at peak periods the area could suffer from noise and air pollution a river runs through the square restricting development and could cause flooding.
7.		For full marks two factors should be mentioned.	6	eg If **New Technology** chosen: **Developing Country** – Increased crop yields better profits for some farmers which can then be used to improve overall standard of living Less physical work for people but fewer jobs for people Expense of machines machines need repairs costing more money which many developing countries cannot afford. **Developed Country** – Increases the efficiency on a farm enabling the farmer to plough, sow, spray etc, more quickly, covering larger areas. It also speeds up harvesting and results in the product being delivered to markets fresher and at a higher premium. It also allows for a smaller work force and therefore lower wage bills. It allows for the use of satellite technology/computers to control the application of fertilisers to particular areas of fields improving yields yet decrease the cost and waste as only the required amounts are delivered to each segment according to the soil quality there.
8.		Answer should be explanation.	6	There is a higher proportion of the population of Kenya under the age of 15 because birth rates are higher in developing countries where fewer women get the chance of an education and there is less use of contraception and information on birth control is less easily accessed. There is a larger number of over 60s in the USA because more people have the chance of medical treatment than in Kenya where there is less money to set up health centres and hospitals. The high living standards in the USA mean that there are plenty of opportunities for women to have careers and this reduces birth rates. Child mortality rates are higher in Kenya so people have many children in order to ensure that some survive. Many families in Kenya have lots of children so they can contribute to the family income when old enough and look after them in old age.

Section 3: Global Issues

Section 3		General Marking Instructions for this type of question	Max mark	Specific Marking Instructions for this question
9.	(a)	Candidates must make a number of relevant, factual points. These should be key points. The points do not need to be in any particular order. Candidates may provide a number of straightforward points or a smaller number of developed points, or a combination of these.	4	The graph shows a steady increase in average global temperature since 1860 of about 1 degree C; there are many fluctuations but from around 1945 to 1965 there was very little overall increase in temperature; since 1970 however, the increase has been faster.
	(b)	Candidates must make a number of points that make the process/situation plain or clear, for example by showing connections between factors or causal relationships between events or processes. These should be key reasons and may include theoretical ideas. Candidates may provide a number of straightforward reasons or a smaller number of developed reasons, or a combination of these. In the case of explaining the formation of a landscape feature, candidates may produce fully labelled diagrams which may warrant full marks if sufficiently accurate and detailed.	6	Answers must refer to both human and environmental effects and should include named examples. A good answer might include explanation and may refer to more than one real life example. Possible answers might include: Warmer temperatures will lead to sea level rise because of melting ice caps, but also because of the thermal expansion of sea water; this is likely to cause significant flooding in many areas such as the Fenlands of eastern England and the Ganges Delta in Bangladesh; in some places such as the Kiribati Islands people have been evacuated because of the danger of flooding. In other parts of the world such as the Sahel area of North Africa, climate change may result in more frequent droughts, affecting people's ability to grow food and leading to more frequent famines. In the UK, some crops such as seed potatoes may not grow as well because of warmer and wetter conditions but farmers may be able to grow different crops such as soft fruit in their place. Warmer global temperatures could also change the habitats of many different species of wildlife, causing a rise in the types of insect pest in the UK for example.
10.	(a)		4	The tropical rainforest climate is much wetter and much warmer than the tundra climate; the wettest month in the rainforest has about 275 millimetres of rain whereas the wettest month in the tundra has under 50 millimetres; the highest temperature in the rainforest is about 27 degrees C but in the tundra it is only 12 degrees C; there doesn't seem to be any seasons in the rainforest but there are very clear seasonal differences in the tundra; the range of temperature in the tundra is 39 degrees C but only 2 degrees C in the rainforest.

	(b)	Answers must refer to both human **and** environmental impacts and should include named examples.	6	eg **tropical rainforest** The Rondonia region of Brazil has suffered from a great deal of deforestation in recent decades, apart from the loss of trees, the habitats of wildlife and the lands of indigenous people such as the Surui have been destroyed; the Surui hunt in the forest and also gather fruits and berries and so not only lose their lands and homes but their food source and their way of life; but they might also get jobs felling trees; once the trees have been cleared, there is less protection for the soil and heavy rain can lead to rapid soil erosion; minerals are leached out of the soil and the soil quickly becomes infertile and useless; deforestation happens not only because of logging companies who export the hardwoods but also because of small scale farmers who clear the forest to try and farm the land; wealthy land owners have cut down vast tracts of forest to establish plantations for soya beans, palm oil and biofuels. eg **tundra** In the Prudhoe Bay area of Alaska, oil drilling and the construction of the Trans-Alaskan pipeline have caused damage to the fragile tundra vegetation and wildlife; local Inuit people have had their way of life disrupted as they are no longer able to access all of their traditional hunting grounds; very few indigenous people have gained employment in the oil industry but some have been adversely affected by contact with western culture, resulting in alcoholism or drug misuse; the oil industry has brought great wealth to the people of Alaska who have the highest average income in the USA. In coastal areas of Greenland climate change is adversely affecting the habitat of animals such as the polar bear and also the traditions of the Inuit as the sea is frozen for a shorter time and the hunting season for both polar bears and people is limited.
11.	(a)	Descriptions must be detailed.	4	Tropical storms are found in areas with warm seas above 26 degrees C; they occur in the Gulf of Mexico and the Caribbean; they are also found in areas of the eastern Pacific such as the South China Sea affecting countries such as the Philippines and Japan; in the Indian Ocean and the Bay of Bengal, cyclones often cause damage to countries such as Bangladesh, India and Sri Lanka.
	(b)		6	In the Ercis and Van region of Eastern Turkey in October 2011, immediate aid was needed to help hundreds of trapped people;emergency rescue teams came from all over the world, often with sniffer dogs and infra-red cameras, to help victims trapped in collapsed building; tents and blankets were needed for the thousands of people who had lost their homes and would have been at risk from hypothermia if they had had to spend the night outside; bottled water was needed quickly as water supplies were cut off; medical help was needed for thousands of injured people. Money would be needed to help rebuild people's houses and also the services they need such as schools, hospitals and clinics; aid from foreign governments or the UN would be used to rebuild important infrastructure such as water pipes, electricity supplies, roads and bridges; training could be given to improve evacuation procedures/earthquake drills; farmers would need to be provided with seeds and tractors to allow crops to be grown to feed people in the future.
12.	(a)		4	Possible reasons: It exports mainly raw materials (primary goods) which are low value goods and imports mainly manufactured goods which are high value. It is dependent on agricultural products for about 34% of its income. It doesn't export many manufactured goods. This country suffers from a trade deficit.
	(b)		6	Possible answers: More money goes directly to the farmer, it cuts out the middlemen who cream off some of the profits. Farmers receive a guaranteed minimum price so they are not affected as much by price fluctuations and can receive some money in advance so they don't run short. More of the money goes to the communities who can invest it in improving their living conditions. Money can be used to provide electricity and drinking water or pay for education. Fair trade also encourages farmers to treat their workers well and to look after the environment. Often fair trade farmers are also organic farmers who do not use chemicals on their crops.

13.	(a)		4	Africa's numbers are low but they show almost a fourfold increase from 1995 to 2020; figures for the Americas and East Asia/Pacific are very similar in 2010, but the latter is expected to increase much faster by 2020 with over 100 million more tourists than the Americas; tourist numbers in Europe are expected to more than double from 1995 to 2020. Middle East tourist numbers are increasing even faster than Europe with the total number expected to almost double between 2010 and 2020.
	(b)		6	For full marks, both benefits and problems must be mentioned. If not, mark out of 4. **Benefits:** Brings in much needed money which should improve the standard of living of the people. Will provide jobs for locals which don't require a lot of skills. Local handicraft industries will also benefit, local people should benefit from improved provision of services. **Problems:** Development of tourist resorts means a loss of farm land. Fishermen lose coastal sites to hotels. Traditional village occupations are decreasing. Increased tourism means an increased demand for water causing water shortages. Beaches can become contaminated with sewage. Local wildlife could be under threat. Employment is only seasonal. Developing countries could become dependent on revenue from tourism and a decrease in tourism could result in an increase in unemployment and cause the economy to decline.
14.	(a)		4	Many developed countries have a very high life expectancy of over 80 years. These include Canada, Japan and Australia. China, some North African countries and Brazil have a life expectancy of 70-79 years. Some of the lowest life expectancies on the planet are found in Africa, particularly in Ethiopia, Congo and Angola where life expectancy is under 50. The lowest life expectancies of all are in some southern African countries where it is under 40 years. Generally, developed countries have a significantly higher life expectancy than developing countries.
	(b)		6	eg **heart disease:** In developed countries such as the UK there is a higher incidence of heart disease because of factors associated with people's lifestyle; often people do not take enough exercise which would help to keep their heart healthy; instead of walking short distances people often take the car for convenience, but walking might be the healthier option; poor diet can also lead to an increased risk of heart disease; for example if there is too much fat in people's diet this could lead to their arteries becoming blocked; not eating enough fruit and vegetables and too much processed food can lead to an increased risk of heart disease; high levels of stress are also linked with heart disease; smoking can also increase the risk. eg **asthma** Asthma can be triggered by other infections such as colds or the flu which can affect the lungs/narrow the airways; it can result from allergic reactions to dust mites indoors or to pollen from certain plants such as oil seed rape; poor air quality can lead to asthma attacks and can result from too many traffic fumes in busy towns/cities; or from cigarette smoke indoors; mouldy or damp conditions in houses can set off an asthma attack too. eg **malaria:** In developing countries malaria is likely to be found in areas where the female anopheles mosquito lives; especially in areas where there is stagnant water for the mosquitoes to breed in; the anopheles mosquito lives in warm and humid areas; lots of people living in close proximity to each other means the mosquito can spread the disease more quickly by biting an infected person and passing on the parasites easily to other people; in shanty towns for example there may be lots of pools of stagnant waste water; also as the population increases there are more rice paddy fields which can be an ideal breeding ground for the anopheles mosquito. eg **kwashiorkor** Kwashiorkor is the result of malnutrition where there is insufficient protein in the diet; it often affects young children when they are weaned from their mother's milk and put on to a diet of mainly starchy vegetables; this might be because families have insufficient knowledge of proper diet or because they live in very poor areas where foods with enough protein are difficult/too expensive to buy; children in communities in tropical/subtropical Africa may be at risk of kwashiorkor, especially where there is already famine which results in poor diet; large family sizes can also be a factor in kwashiorkor where there is not enough of the right types of food to go round.

NATIONAL 5 GEOGRAPHY MODEL PAPER 1

Section 1 – Physical Environments

1 (a) U-shaped valley – 979976
    Corrie – 957997
    Pyramidal peak – 954976   3

   (b) 1 mark for a valid point.
       2 marks for a developed point.
       Full marks can be gained for appropriately annotated diagrams.
          If a U-shaped valley is chosen: a glacier occupies a V-shaped valley (1), the ice moves, eroding the sides and bottom of the valley (1), through plucking (1) and abrasion (1); this makes the valley sides steeper (1) and the valley deeper (1), and when the glacier retreats a steep, deep, flat-floored U-shaped valley is left (2), the original stream seems too small for the wider valley and is known as a misfit stream (1).
                                                                  4

2 (a) Pot hole – 833164
    Intermittent drainage – 817155
    Limestone pavement – 814175   3

   (b) 1 mark for a valid point, 2 marks for a developed point.
       Limestone pavement:
          Limestone made from decayed remains of skeletons of sea creatures (1); laid in horizontal layers on sea bed (1); sedimentary rocks uplifted (1); overlaying rock removed by glaciation (1); cracks appear as rock dries out (1); cracks widen into grykes by chemical weathering (1); limestone dissolved by acid rainwater (1); creates clints, or upstanding blocks (1).
       Credit should be given for appropriately annotated diagrams.   4

3 1 mark for a simple point.
   2 marks for a developed point.
   Max 1 for grid references.
   Answers could include:
       Most settlement is found along the course of the river where the land is low and suitable for building (1), e.g. at 885125 (1). Transport routes such as the B970 and the railway line follow the course of the valley as it provides a natural route way (2). The roads have had to avoid the higher, steeper ground and in some places the undulating nature of the land means cuttings and embankments are needed (1) such as at 892132 (1). Where there is gently sloping lower ground (901117) farming may be arable or mixed (1), whereas on the steeper slopes and higher ground livestock grazing will occur due to the difficulty of using machinery (2). Woodland is grown on land that is too high and cold for crops, such as 905066 (1). Steep slopes are no use for settlement but can be for recreation and tourism, such as the ski centre at 899059 (1).
   Accept any other relevant point.   5

4 Max 2 marks for description of weather changes without explanation.
   Mark out of 3 if candidate has misidentified fronts but explained weather correctly.
   Answers could include:
       As the warm front approaches Norwich, air pressure will fall (1), cloud cover will increase (1) and steady rain will occur (1). Winds will be quite strong as the isobars are close together (1). The warm front will move away, and Norwich will be in the warm sector of a low-pressure system (1). Temperatures will rise, and it will be mild with occasional showers and some cloud cover (1). Winds will die down. The cold front will arrive and cloud cover will increase, with cumulonimbus clouds bringing heavy rain to the city (2). Temperatures will drop as the cold front passes over and begins to move away (1). The sky will become clear (1), the rain will stop (1) and pressure will begin to rise (1), and winds will increase (1).
   Accept any other valid point.   4

5 (a) 1 mark for a valid point.
       2 marks for a developed point.
       If glaciated uplands chosen:
       • Farming – Highlands are suitable for sheep farming as opposed to arable farming (1). Sheep can survive on steep slopes (1). Steep land is unsuited to machinery used on arable farms (1).
       • Forestry – Trees can be planted on the upper slopes where the land is unsuitable for other crops due to weather and steepness of slopes (1).
       • Recreation and tourism – Tourists are attracted to mountain scenery (1). The steep slopes can be used for skiing (1). The mountains attract hill walkers (1).
       • Water storage and supply – Valleys can be dammed to create reservoirs (1). Mountain areas usually have high rainfall (1). The underlying rock may be impermeable and therefore suitable for water catchment and storage (1).
       • Renewable energy – Highlands provide suitable sites for HEP schemes (1). Water flowing down pipelines can be used to drive turbines to create electricity (1).
       • Mountain areas are also suitable sites for wind farms because of the frequency and strength of the local winds (1).   4

Section 2 – Human Environments

6 (a) 1 mark per valid point.
   Answers could include:
   Main roads like the A691 and the A690 converge on this square (1); there is a tourist information centre (1) and a cathedral (1), many churches (1) and a bus station (1).
   Accept any other valid point.   3

   (b) 1 mark per valid point.
   2 marks for a developed point. Mark only differences.
   Answers may include:
   2745 is an area of newer housing, 2642 is an older, inner urban area (1); 2745 is mainly a residential suburban area whereas 2642 has a greater variety of land uses like industry and services, as well as housing (2). 2745 has a varied street pattern including cul-de-sacs and crescents whereas 2642 has a mainly rectangular/grid-iron pattern (1); 2745 has mainly small buildings (houses) whereas 2642 has large buildings like factories (1); 2745 has a limited amount of traffic and less noise pollution, while 2642 has many main roads, railway, bus station – more noise and pollution (2).
   Accept any other valid point.   5

7 1 mark for a valid point.
   2 marks for a developed point.
   If no named example, mark out of 5.
   If the Rio de Janeiro example was chosen, answers may include:
   New roads are to be constructed to improve the transport of people and goods in the area (1); storm drains are to be built to control flooding (1); construction of piped water supply (1); construction of latrines (1). Slums will be cleared over a five-year period (1), and people are being rehoused nearby in newly built apartments (1). These are affordable accommodation (1). The estates also include schools, markets and other facilities (1).
   Or any other valid point.   6

8 For full marks, two factors should be mentioned.
   Award up to 4 marks for any single factor.
   1 mark per valid point.
   2 marks for a developed point.
   If New Technology is chosen, answers may include:
   Machinery increases the efficiency on a farm, enabling the farmer to plough, sow, spray etc. more quickly, covering larger areas (2). It also speeds up harvesting and results in the product being delivered to markets fresher (1) and at a higher premium (1). It also allows for a smaller work force (1) and therefore lower wage bills (1). It allows for the use of satellite technology/computers to control the application of fertilisers to particular areas of fields (1), improving yields (1) yet decreasing the cost and waste.
   If Diversification is chosen, answers may include:
   Farmers can obtain additional income from a variety of sources if they diversify their activities on the farm (1). They may turn old farm workers' cottages into holiday chalets (1). They may use part of the land for a golf course (1). They may earn income from sports such as quad-bike riding (1).
   Or any other valid point.   6

Section 3 – Global Issues

9 (a) Answers may include:
   Over a 17-year period between 1979 and 1996, ice melt decreased (1), apart from 1983–4 when it increased (1). The general trend for 1997 onwards was that ice melt increased (1), reaching a high in 2012 (1) at 2% higher than the average (1).
   Or any other valid point.   4

   (b) 1 mark for a valid point.
   2 marks for a developed point.
   Answers may include:
   Warmer global temperatures could change the habitats of different wildlife and marine dwellers (1). Warmer water causes some fish to die or to move to colder waters, which affects fishermen and their livelihood (1). Increased temperatures may mean more drought, leading to famine (1). In some areas, new crops can be grown because of the higher temperatures (1), which increases farmers' income (1). Rising temperatures can cause the icecaps to melt, which can result in flooding in low-lying areas like the Netherlands (1). Warmer temperatures could result in the spread of diseases like malaria into new areas (1).
   Or any other valid point.   6

10 (a) Answers may include:
   The lowest temperature reached in Barrow is –29 degrees whereas in Eismitte it is –27 degrees Centigrade (1). The highest temperature reached in Barrow is 5 degrees, compared with 12 degrees Centigrade in Eismitte (1). The range in temperature in Barrow is therefore 34 degrees as opposed to 39 degrees in Eismitte (1). Barrow has precipitation throughout the year, totalling 130mm, whereas there is no rainfall in Eismitte (2).
   Or any other valid point.   4

   (b) Answers may include:
   Large areas of rainforest are needed for cattle ranching, so huge areas are burned, destroying the native wildlife and the plants (1). The land the cattle graze soon becomes infertile, so more rainforest is destroyed to allow the cattle to graze (1). This land never recovers (1). Large tracts of forest are cleared by using fires, which can get out of control and destroy far more forest than necessary (1). Large amounts of CO_2 are released, affecting the local climate as well as the global climate (1). With no trees to bind the soil together, soil erosion takes place (1). Rivers become polluted with soil and cannot be used by the local people (1). The local people are forced off the land, and their traditional way of life is under threat (1). Logging destroys the habitats of animals, causing a threat of extinction (1). Fewer animals means less food for the native people (1). Illegal loggers do not replant to replace the trees, so the rainforest does not regenerate (1).
   Or any other valid point.   6

11 (a) Answers may include:

Volcanoes are found along the edges of plate boundaries (1). Many volcanoes are found along the west coast of North and South America (1) especially Alaska (1). Volcanoes are found along the coastal areas of China and Japan (1) and along the Pacific Ring of Fire (1).

Or any other valid point.   4

(b) 1 mark for a valid point.

2 marks for a developed point.

If no specific example named, mark out of 4.

Answers may include:

Tropical storms are very powerful and can cause immense damage (1). Storms can uproot trees (1) and disrupt telephone lines and electricity power lines (1). If plantation crops are destroyed, this can cause severe economic problems (1). Tidal surges flood low-lying coastal areas (1). There can be major loss of life caused by flooding, resulting in thousands of deaths (1). Landslides may occur where rainfall washes away buildings built on steep, unstable slopes (1). Flooding can block coastal escape routes and relief roads, making damaged areas unreachable for rescue services (2).

Or any other valid point.   6

12 (a) Answers may include:

The percentage share of world goods production for developed countries will steadily decrease between 2000 and 2016 (1). The percentage will have decreased from 68% to 45% in developed countries – a drop of 23% (1). China's share will steadily increase during this period from 4% to an estimated 18% (1). India's percentage has also increased, but the increase is significantly lower than that of China (1). India's share remained at 4% between 2000 and 2010 but is due to increase slightly to 7% by 2016 (1).

Or any other valid point.   4

(b) 1 mark for a valid point.

2 marks for a developed point.

Answers may include:

Developed countries have a larger share of world trade because their exports include significantly more manufactured goods than countries in the developing world (1). Developing countries tend to produce raw materials rather than manufactured goods (1). Developed countries have more industries producing a wide variety of products, such as food products, industrial machinery and electronics, which are traded with other developed countries (2). Many developed countries belong to trading alliances, such as the European Union, which help to increase the volume of trade (1). The economies of developed countries benefit from being able to purchase low-cost raw materials produced by developing countries and to sell manufactured goods back for higher profits (2). Developing countries have much less money to invest in manufacturing industries and are less able to compete with developed countries (1).

Or any other valid point.   6

13 (a) 1 mark for a valid point.

2 marks for a developed point.

Answers may include:

The overall number of tourist arrivals across the world has steadily increased throughout the period from 1990 to 2011 (1). The number has more than doubled from 1990 to 2011 (1). The highest number of arrivals in 2011 occurred in Europe, with France being the most visited country (2). The second largest number of arrivals occurred in Asia, Australia and Oceania, with China accounting for 57·6 million visitors (2). North and South America were third largest in tourist arrivals, with the USA having 62·3 million tourist visitors (2). Africa had the smallest share of tourist arrivals with only 5%, as opposed to 51·5% in Europe (2).

Or any other valid point.   4

(b) 1 mark for a valid point.

2 marks for a developed point.

For full marks, both advantages and disadvantages must be mentioned. If not, mark out of 5.

Advantages:

Mass tourism boosts the economies of developed and developing countries (1). Tourism creates jobs in a wide variety of activities, including farming (supplying food for shops/hotels etc.) and entertainment (theatres, leisure centres) (2). It provides money for the country to improve infrastructure (transport, water supplies, sewage disposal) (1) and provides income to improve services such as education and health services (1).

Disadvantages:

Tourism can create pollution, e.g. litter on beaches (1). Sea and river pollution arises from the increased use of fertilisers and pesticides on local farms (1). Can increase traffic congestion in both cities and rural areas (1). Can cause conflict in rural areas with farmers to produce more food to feed tourists (1). Employment may only be seasonal (1). Tourism can have a detrimental effect on local culture and the physical environment, for example removing forest to build more tourist facilities (1).

Or any other valid point.   6

14 (a) 1 mark for a valid point.

2 marks for a developed point.

Answers may include:

AIDS is most prevalent in countries in the developing world (1). The highest percentages of infected adults are found in areas such as Central and South Africa (1), which have 15 to 34% of the population infected (1). Infection rates are much lower in areas such as Europe and North America at 0·1 to 0·5% (1). Infection rates are also low in Australia (1). Rates are also high in parts of Asia, particularly in India and Pakistan (2).

Or any other valid point.   4

(b) (i) 1 mark for a valid point. 2 marks for a developed point.

Max 1 mark for a list.

If heart disease is chosen, answers may include:

Heart disease can be inherited from parents (1). Over-eating can lead to obesity, putting extra pressure on the heart (2). Smoking narrows the arteries and affects the lungs (1). This can lead to emphysema, putting a strain on the heart by making it work faster (1). The build-up of fatty deposits on the walls of the arteries

restricts the flow of blood to the heart (1); lack of exercise raises blood pressure, affecting the efficiency of the heart (1); poor diet increases cholesterol (1); stress leads to high blood pressure (1).
Or any other valid point. 3

(b) (ii) Max 1 mark for a list.
Answers may include:
More people now have regular check-ups for cholesterol and blood pressure (1), allowing early intervention for at-risk patients (1). More advanced medical equipment is being invented and used (1), e.g. artificial heart valves (1). More advanced surgery is now available (1), e.g. bypass surgery (1). The government runs many campaigns to educate the public (1), e.g. stop-smoking campaigns and healthy eating (1). Advertising brands of cigarettes has been banned (1). Smoking ban in public places (1). Nicotine patches to help people give up smoking (1). People are encouraged to eat more healthily and take more exercise (1). Healthy eating is encouraged in school dining halls (1). Free and reduced membership of gyms (1).
Or any other valid point. 3

NATIONAL 5 GEOGRAPHY MODEL PAPER 2

Section 1 – Physical Environments

1 (a) Stack – 636871
Headland – 570863
Bay – 555869 3

(b) 1 mark for a valid point.
2 marks for a developed point.
- Caves – Caves occur where a coastline of hard rock is attacked by prolonged waves (1). The waves attack along a line of weakness such as a joint or fault in the rock (2).
- Arches – Over time, erosion of the cave may cut through the headland (1). When it reaches the other side, an arch may be formed (1).
- Stack – Continual erosion at the foot of the arch may cause the roof to collapse (1). This creates a piece of rock isolated from the headland called a stack (1).
Or any other valid point. 4

2 (a) Confluence – 584973
Meander – 614965
River flowing south – 670967 3

(b) 1 mark for a valid point.
2 marks for a developed point.
A fully annotated diagram can obtain full marks.
Waterfall – Waterfalls occur in a river valley at a point where soft rock is overlain with harder rock (1). Water flowing over the hard rock begins to erode the softer rock underneath (1). At the point of erosion, a pool called a plunge pool is formed (1). As the softer rock wears away, it can no longer support the harder rock above, which collapses into the plunge pool (1).
Or any other valid point. 4

3 1 mark per valid point.
2 marks for an extended point.
1 mark for a grid reference.
Answers may include:
There is woodland for forest walks (1) at 533903 (1). The coastline attracts tourists for the beautiful views (1). The limestone cliffs are attractive for rock-climbing (1). Oxwich Bay is attractive for its beautiful beach and sand dunes (2). Caves like Bacon Hole (560867) along the coast attract cavers (2). The sea can be used for water activities like boating, surfing and swimming (1).
Accept any other valid point. 5

4 For full marks, both similarities and differences must be mentioned.
Otherwise mark out of 3.
Similarities:
There are few clouds in the sky (1). The weather is dry (1). Wind speeds are light and calm towards the centre of the anticyclone (1). There may be mist in the morning at both times of the year (1).
Differences:
Temperatures are high in summer and low in winter (1). During the night in summer, temperatures remain above zero (1), but in winter they will fall below zero. Warm moist air rising from the ground can form thunderstorms in summer (1). Heatwaves can occur in summer if the anticyclone remains for a period of time (1).
Or any other valid point. 4

5 1 mark for a valid statement.

2 marks for a developed point.

If Farming/Recreation and Tourism is chosen, answers may include:

> Walkers climbing over stone walls could damage them (1). Sheep may be frightened by walkers' dogs (1). Gates could be left open, allowing animals to escape (1). Tourists could be denied access to areas (1). Tourists could cause an increase in traffic congestion on small country roads (1). Tourists could drop litter which animals could eat (1). Farmers could cause some river pollution through fertilisers and pesticides draining into rivers (1).

Or any other valid point.   4

Section 2 — Human Environments

6 1 mark per valid point.

2 marks for an extended point.

For full marks, both advantages and disadvantages should be mentioned. If not, mark out of 4.

Advantages:

> There is flat land suitable for building on (1). There is room for expansion (1). Swansea is close by, providing a market and a labour force (2). The B4620 allows transport of goods in and out as well as easy access for customers (2). Many people will travel by car, and there is room for parking (1). The site is on the outskirts of town, so the land will be cheaper (1) and traffic congestion less likely than a location in the town centre (1).

Disadvantages:

> There are electricity transmission lines taking up space in the square, limiting the available area for building (1). The Afon Llan runs through the square, which could cause flooding (1). There is only a B-class road running into the square, so it may not be able to handle the amount of traffic and the size of vehicles (2). There is already an industrial estate in the square, so traffic congestion may occur (1).

Accept any other valid point.   5

7 (a) 1 mark for a valid point.

2 marks for a developed point.

Answers may include:

> The most densely populated areas include Europe, the east coast and parts of the west coast of the USA, the east coast of South America, north-west Africa, coastal India and the eastern coasts of China (2).

> Moderately populated areas include the Midwest and western parts of the USA, Scandinavia and eastern and central Russia, coastal Australia and central and southern Africa (2).

> Sparsely populated areas include northern Canada, central South America, central Australia and northern Africa (2).

Or any other valid point.   3

(b) 1 mark for a valid point.

2 marks for a developed point.

For full marks, both physical and human factors should be mentioned. Otherwise, mark out of 4.

Answers may include:

> Physical factors such as relief, high mountains and climate, desert areas and polar regions limit population (2). Many parts of the world are inaccessible, such as tundra areas (1). People are attracted to areas of flat land such as river valleys and coastal areas because land is easier to build on (2). Soils are usually more fertile in these areas for farming (1). Areas which have natural resources such as wood, oil, coal and water attract higher numbers of people (1).

> Human factors include employment opportunities; where people can find jobs; and industry (1).

> Governments can offer financial incentives to attract industries which can attract employment into an area (1) and therefore people (1). If areas are easily reached through good transport links, there is likely to be more work available (1).

Or any other valid point.   6

8 1 mark for a valid point.

2 marks for a developed point.

If exports are chosen, answers may include:

The USA is more developed than Chad because it exports manufactured goods, which leads to more money coming into the country (1), whereas Chad exports mainly raw materials, which bring in much less money than manufactured goods (1). The USA will have higher levels of pay and employment than Chad, and therefore the standard of living for the population in the USA will be much higher than in Chad, where wages will be low (1).

If literacy rates are chosen, answers may include:

Literacy rates are much higher in the USA than in Chad because better standards of education are provided by the government (1). There will be more schools, colleges and universities available to educate people in the USA (1). Since 99% of people in the USA can read as opposed to 35% in Chad, people will have a better chance of obtaining employment in a wide range of jobs in the USA (1).

Or any other valid point. 6

Section 3 – Global Issues

9 (a) 1 mark for a valid point.

2 marks for a developed point.

Answers may include:

Average global temperatures have steadily risen from 13·7 degrees in 1880 to a predicted 14·5 degrees in 2020 (1). Throughout that period, the temperature has fluctuated over various 20-year periods (1). From 1880 to 1920, there were four times when the temperature fell below 13·7 (1). From 1920 to 1960, the temperature reached a high of about 14·3 in 1940 (1). Again there were several times when the temperature fell during this period, such as in 1960, when it fell to 13·8 degrees (1). In the period 1960 to 2000, the highest temperature reached was in 2000, when it rose to about 14·6 degrees (1). The lowest temperature in this period was about 13·8 degrees around 1970 (1).

Or any other valid point. 4

(b) Both physical and human causes must be mentioned for full marks. Otherwise, mark out of 4.

1 mark for a valid point.

2 marks for a developed point.

Physical causes:

These include times when there were sun flares, which helped to increase temperatures (1). Gases such as methane given off from rotting vegetation in areas such as the tundra cause a natural increase in air pollution, again causing temperatures to rise (1). These changes in temperature can cause winds such as the Westerlies to change course (1). These can cause changes to the climates in different areas of the world which are affected by offshore winds (1). The temperature changes in the seas and oceans can cause changes in ocean currents (1). These can have an adverse effect on the climates of coastal areas through flooding and hurricanes (1).

Human causes:

The burning of forests to clear land for other human activities releases vast amounts of carbon dioxide into the atmosphere, which pollutes it (1). Emissions from road-transport exhaust fumes also release toxic pollution into the atmosphere (1). Industrial plants release additional gases which pollute the atmosphere (1). Some scientists have suggested that these pollutants have punctured holes in the ozone layer which protects the Earth against harmful rays from the sun (1). The use and testing of atom bombs during the 1940s, 50s and 60s released radioactive particles into the atmosphere and beneath some oceans (1). Gases known as CFCs have been released from waste sites where disused fridges have been dumped (1).

Or any other valid point. 6

10 (a) 1 mark per valid point.

2 marks for a developed point.

Answers may include:

There was a fall in the number of square miles of deforestation from 8000 to 4500 between 1988 and 1990 (1). From 1990 to 1992, deforestation increased to a peak of 11000 square miles (1). This was followed by a further drop to just under 6000 square miles by 1996 (1). From 1996 to about 2003, deforestation rose steadily to a peak of 10000 square miles (1). Since then, there has been a continual drop to just over 2000 square miles in 2010 (1).

Or any other valid point. 4

(b) 1 mark for a valid point.
   2 marks for a developed point
   Answers may include:
      National parks have been set up in rainforest
      areas to protect the land and people from the
      environmental effects of mining (river pollution),
      cattle ranching and road building (2).

      In places such as Colombia, the government has
      returned land to local people which had been taken
      away from them (1). Developed countries are less
      inclined to fund projects which are harmful to the
      rainforests (1). Laws against illegal logging have
      been passed to reduce deforestation (1). To protect
      the environment, logging companies have been
      encouraged to replant trees in areas where trees
      have been removed (1). Agro-forestry schemes have
      been introduced whereby forests are used for small-
      scale farming schemes (1). These provide sustainable
      crop yields and protect the ways of life of local
      people (1).
      Or any other valid point.   6

11 (a) 1 mark for a valid point.
   2 marks for a developed point.
   Answers may include:
      Hurricanes occur in the Caribbean in areas such as
      Jamaica (1). They also occur along the south-east
      coast of the USA, affecting areas such as Florida
      (1). They occur in an area stretching from Oceania
      to the south-east coast of Africa (1). Typhoons are
      found in South-East Asia stretching across the Indian
      Ocean to coastal India and Bangladesh (2).
      Or any other valid point.   4

(b) 1 mark for a valid point.
   2 marks for a developed point.
   For earthquakes, answers may include:
      Scientists can predict where earthquakes will occur,
      but it is almost impossible to predict when (1).
      Seismograph readings do not give enough warning (1).
      Despite advanced technology, extensive damage and
      loss of life can happen in built-up areas such as Los
      Angeles or Kobe (2).
      It is better to build earthquake-proof buildings in
      known areas where earthquakes occur (1). It is also
      better to plan for an earthquake by having rescue
      services and medical services trained and ready to
      deal with a disaster (1).
   For volcanic eruptions, answers may include:
      Tremors and gas emissions give warning to local
      people (1). This allows time for evacuation (1).
      Although the landscape can be destroyed, evacuation
      prevents huge loss of life (1). If a volcano is becoming
      active, it may expel lava bombs which will warn the
      population to leave the area before it erupts (1).
      Areas where volcanoes are active can plan to have
      rescue/aid measures in place to reduce the impact of
      an eruption, such as the destruction of buildings and
      roads by flowing lava (1).
   For tropical storms, answers may include:
      Satellite images can provide plenty of warning (1),
      allowing the population time to evacuate the area
      (1) and protect buildings (1). People can prepare to
      protect themselves by building storm shelters under
      the ground (1). Storm warnings can be given over
      the radio or television (1). Local people often stock
      up on food and water supplies to use after the storm
      has passed (1).
      Or any other valid points.   6

12 (a) 1 mark for a valid point.
   2 marks for a developed point.
   Note that it is not necessary to describe the changes
   for all four countries for full marks.
   Answers may include:
      From 1972 to about 1998, the United States'
      percentage share of world exports fluctuated
      between 10 and 12%, having begun the 1970s at
      about 14% (2). From 1998, the USA's share fell to
      just over 8% (1). Germany's share was at its highest
      in the later part of the 1980s with a peak of just
      over 12% (1). Since then, it has gradually fallen
      to about 9% (1). Japan's share increased from
      6% in about 1972 to a high of 10% in around 1985
      (1). Since then, it has decreased to just under 5%
      in about 2008 (1). China's percentage share rose
      steadily from about 0·7% in 1970 to just over 3% in
      about 1997 (1). It then increased dramatically to
      over 12% by 2010 (1).
   Or any other valid point.   4

(b) 1 mark for a valid point.
   2 marks for a developed point.
   Candidates can obtain additional marks by referring
   to a named developing country or countries they may
   have studied. For full marks both people and the
   environment should be mentioned.
   Answers may include:
      Some countries are too dependent on just one
      or two exports (1). Ghana, for example, depends
      on cocoa for 85% of its exports (1). If the price of
      cocoa were to fall, cocoa farmers would receive
      less income (1). They would have less to spend,
      affecting other businesses in Ghana (1). If the
      price of the export in a developing country was to
      increase, then there would be more income for the
      country to invest in other products and industries
      (1). Countries would be less affected if they had a
      greater range of products to export (1). This could
      lead to improvements in infrastructure, education
      and health (2).
      Or any other valid point.   6

13 (a) 1 mark for a valid point.
   2 marks for a developed point.
   Answers must refer to differences. It is not necessary
   to refer to all of the parks for full marks.
   Answers may include:
      The most popular national park is the Lake District
      with 15·8 million visitors a year compared with 4
      million visitors in Loch Lomond (1). There is also
      a large difference in the income for these two
      parks, with Loch Lomond earning £150 million as
      opposed to the Lake District's £952 million (1). The
      Cairngorms earn £185 million from their 1·5 million
      visitors, whereas the other mountainous area,
      Snowdonia, obtains a much larger income of £396
      million from a total 4·7 million visitors (1).
      Or any other valid point.   4

(b) 1 mark for a valid point.
   2 marks for a developed point.
   If no area is named, mark out of 4.
   Answers may include:
      In coastal areas such as Dorset, the government
      has tried to encourage more tourists to visit inland
      areas in order to reduce pressure on coastal areas
      (1). There are laws forbidding the dropping of
      litter, and fines are used to enforce them (1). Under
      EU regulations, clean beaches can be awarded a

Blue Flag (1). A number of scenic areas in Britain have been designated as National Parks, such as the Lake District and Loch Lomond (1). Future developments such as tourist hot spots can be monitored and controlled (1). Areas within the parks can be designated for car parks and other buildings (1). In mountain areas, money can be spent on improving footpaths which have been eroded over time by hillwalkers (1). Visitors to these areas can be educated to respect the natural environment through information centres (1). Sports activities such as mountain biking and quad biking can be restricted to certain areas (1). There are organisations such as the National Trust which make great efforts to protect the environment, culture and historic buildings in popular tourist areas (1).
Or any other valid points. 6

14 (a) 1 mark for a valid point.
2 marks for a developed point.
Answers may include:
Cholera outbreaks have been reported throughout the world but particularly in developing countries (1). Apart from Europe, outbreaks have been reported in central and southern Africa, India and islands in South-East Asia, islands in the Caribbean and parts of Oceania (2). In places such as the USA, Britain, Germany and Australia, there have been cases where people travelling from other countries have imported some cases of cholera (1).
Or any other valid point. 4

(b) 1 mark for a valid point.
2 marks for a developed point.
Answers must refer to both causes and prevention methods for full marks. Otherwise, award a maximum of 4 marks.
For malaria, answers may include:
Malaria is spread by the female anopheles mosquito (1). The disease is carried by the mosquitoes, which have been infected by taking blood meals from infected persons and then pass it on in their saliva when biting another person (2). The mosquitoes breed in areas of still water such as swamps or even in water barrels (1).
Methods to prevent/reduce the disease include the use of insecticides to kill mosquitoes (1); using anti-malarial drugs such as chloroquine to treat blood parasites (2); releasing water from dams to drown immature larvae (1); draining breeding grounds such as swamps (1); using small fish to eat larvae (1); using mustard seeds to drag larvae below the surface to drown them (1); using nets to protect people from being bitten by mosquitoes when sleeping (1).
Or any other valid point. 6

NATIONAL 5 GEOGRAPHY MODEL PAPER 3

Section 1 — Physical Environments

1 (a) Arête – 472693
Corrie – 467677
U-shaped valley – 525594 3

(b) Full marks can be awarded for diagrams which clearly illustrate/explain the formation of a corrie.
Answers may include:
Snow fills a hollow on the side of a mountain and is compressed and turns to ice (1). Ice moves downhill under gravity (1). The corrie floor is abraded (1). The ice plucks rock from the back wall (1). The hollow is deepened by erosion (1). A lip is left as the ice loses power (1). Freeze/thaw action helps to steepen the back wall (1).
Or any other valid point. 4

2 (a) V-shaped valley – 473657
Meander – 447585
Tributary – 528595 3

(b) 1 mark for a valid point.
2 marks for a developed point.
A well-annotated diagram could obtain full marks.
Answers may include:
A V-shaped valley is the upper part of its course which is eroded by the river (1). It is created by down-cutting/vertical erosion (1) by corrosion and hydraulic action (1). The exposed sides are weathered, for example by freeze/thaw action (1). Particles of rock are moved down the slope by the movement of rainwater and gravity (1) and also by being transported away by the fast-flowing stream (1).
Or any other valid point. 4

3 1 mark for a single point.
2 marks for a developed point.
For full marks, reference must be made to at least two of the land uses.
Maximum 1 mark for a grid reference.
Answers could include:
• Deer stalking – On the highest land where it is too cold and the soils are too thin even for trees to grow (2). Deer are nimble and so can cope with steep slopes and can survive on rough grazing (2).
• Forestry – Most of the land is above 250m where it is too cold for crops to grow (1) and the growing season is too short (1). Soils are acidic and rainfall is high, but coniferous trees can grow in these conditions (2). Many of the slopes are too steep to use machinery (1).
• Mixed farming/settlement – Land is lower and the climate is warmer, so is more suited to settlement and cultivation, producing a pattern of scattered farm houses (2). Arable farming can take place on the flat alluvial soils of the Peffer flood plain (1). Livestock can be grazed on the steeper, sloping land (1).
• Ancient fort – A good defensive site on top of a hill (1) from which the advancing enemy could be observed (1). It would have been much easier to repel them from here than from down by the river (1).
Or any other valid point. 5

4 1 mark for a valid point.

2 marks for a developed point.

The answer must explain the differences in weather.

Answers may include:

> There is greater cloud cover in Cape Wrath than in Banbury because a cold front has just passed Cape Wrath whereas Banbury is in the warm sector (1). Wind speeds are higher in Cape Wrath, as shown by the isobars being closer together than in Banbury (1). Temperature is lower in Cape Wrath because of the cold front, whereas Banbury is in the warm sector, which brings higher temperatures (1). The cold north-westerly winds have brought snow and sleet in Cape Wrath, whereas there is little or no rain in Banbury because of the warm sector (1).

Or any other valid point. 4

5 1 mark for a valid point.

2 marks for a developed point.

For glaciated uplands and water storage and supply, answers may include:

> Glaciated uplands in Britain contain lochs and lakes (1). These are used to store water and to supply water to towns and cities (1). Local climate and underlying geology are important, and glaciated uplands are very suitable in this respect (1). These areas are located in areas which have high average rainfall, such as the north and west of Britain (2). The rocks in these areas are mainly impermeable rocks such as granite (1). This allows storage reservoirs to hold water (1).

Or any other valid point. 4

Section 2 – Human Environments

6 1 mark for a valid point.

2 marks for a developed point.

No marks for grid reference.

Accept yes/no answers.

Answers may include:

Advantages:

> The area is flat for building houses (1). There are pleasant views over the water (1). It has good road communications, allowing easy access to the area by car (1). There is also a railway station close by, so residents can travel by train (1). There is a hospital near the site in case of accidents (1). The town of Dingwall is within walking distance for provisions (1). There are leisure activities close to the location, e.g. leisure centre, museum and a castle (546601) (2).

Disadvantages:

> The housing will be close to a rifle range, which could be noisy and dangerous (1). There are works at 560585 which would be an eyesore (1). It is close to the water, so could be dangerous for children (1). The railway line is close by, which could be noisy (1).

Accept any other valid point. 5

7 (a) 1 mark for a valid point.

2 marks for a developed point.

Answers may include:

> The basic link between GDP per capita and the number of births per woman is that the higher the GDP, the lower the number of births per woman (1). The lower the GDP per capita, the higher of number of births per woman (1). For example, Austria has both the highest GDP per capita ($31000) and the lowest births per woman (1), whereas Guinea has the highest births per woman (six) but the lowest GDP per capita ($3000).

Or any other valid point. 3

(b) 1 mark for a valid point.

2 marks for a developed point

The number of births per woman is low where GDP per capita is high because:

> Women have access to family planning and contraception (1). Women are better educated to follow careers (1). This means that they marry later and delay having children until they are older (1). Therefore they have smaller families (1). Many women work instead of staying at home to look after children (1). Infant mortality rates are low, therefore there is less need to have so many children (1).

The number of births per woman is high where GDP per capita is low because:

> There is a lack of birth control and family planning (1). Poverty and lack of health care means that many children die in infancy (1). So, parents have more children in the hope that a few will survive (1). In poor countries, children provide a workforce and earn money for their families (1). A lack of pensions and social benefits means that children are needed to look after their parents in old age (1). Religious and social pressures encourage people to have more children (1). Women marry young and have larger families (1).

Or any other valid point. 6

8 1 mark for a valid point.

2 marks for a developed point.

Developed countries:

As there are fewer children, this affects the numbers attending school, leading to school closures (1). There will be a smaller number of people available for work (1).There will be more elderly people dependent on a smaller number of people to support them (1). More money will be needed to provide medical and social care for an ageing population (1).

Developing countries:

Large families may be forced to live in poverty (1). Less food may be available for the population (1). Less money will be available for medical care and education (1). More children may die in infancy (1). Literacy rates will be much lower than in developed countries (1). Less industry means that unemployment will be high, with large families suffering the effects most (1). Standards of living will be low for many families (1). However large populations can produce a large workforce.

Or any other valid point. **6**

Section 3 — Global Issues

9 (a) Both carbon dioxide and global temperature must be mentioned for full marks. Otherwise, mark out of 3.

Carbon dioxide concentrations:

Carbon dioxide concentrations rose from 300 parts per million in 1900 to almost 400 parts per million in 2010 (1). From 1900 to 1950 there was hardly any change, but since 1950 carbon dioxide concentrations have steadily increased (1).

Temperature changes:

Temperatures continued to increase during the period 1900 to 2010 (1). During this period, global temperatures increased by 1·5 degrees (1). Global temperatures fell in about 1915 and in 1950 by 0·25 and 1 degree respectively (1).

Or any other valid point. **4**

(b) 1 mark for a valid point.

2 marks for a developed point.

At least two ways must be mentioned for full marks. Otherwise, mark out of 5.

Answers may include:

Introducing laws to reduce the burning of forests (1); introducing replanting schemes where forests have been destroyed (1); reducing the use of sprays which include CFCs and ensuring that no CFCs are allowed to escape from fridges etc. on waste dumping sites (1); reducing exhaust emissions containing lead and carbon dioxide by adding filter systems to vehicle exhaust systems and producing cars/lorries which use lead-free fuel (1); reducing the use of fossil fuels such as coal, oil and natural gases by introducing 'green' fuels such as HEP, wind power, solar power and other renewable energy sources (2).

Or any other valid point. **6**

10 (a) 1 mark for a valid point.

2 marks for a developed point.

Answers may include:

The highest percentage increase in forest cover during the period 1985 to 2010 was in Vietnam with +38%, with the lowest percentage increase occurring in India with +6% (2). The second highest increase was in China, with +25% (1). The highest percentage loss was in the Philippines with −32%, and the second lowest percentage loss occurred in Malaysia and Laos, each with 7% (2). Other countries suffering a loss included Myanmar −18%, Indonesia −25%, Cambodia −20% and Thailand −10% (2).

Or any other valid point. **4**

(b) 1 mark for a valid point.

2 marks for a developed point.

If no area is mentioned, mark out of 5.

Advantages:

In Brazil, large areas have been cleared by timber companies, and the hardwood has been exported abroad (1). This earns income for the country (1). Forests have been cleared to make room for new farmland (1) and settlements for the expanding population (1) and to increase food production (1). Forest is also destroyed for mineral extractions which are also sold to other countries (1).

Disadvantages:
The habitats of wildlife are destroyed (1). Burning trees releases vast quantities of carbon dioxide into the atmosphere and may contribute to global warming (1). The homes of indigenous tribes are destroyed, as are their traditional culture and way of life (2). Plants which may contain cures for diseases are also destroyed (2). Poor farmers lose their land and may be forced to migrate to towns and cities to find employment (1).
Or any other valid point.   6

11 (a) 1 mark for a valid point.
2 marks for a developed point.
Answers may include:
Earthquakes occur along or near the boundaries of large plates which make up the earth's crust (2). These areas are known as plate margins (1). These are found in southern Europe, through the Middle East and into eastern and South-East Asia (2). There are also earthquakes stretching from Alaska down through the west coast of the USA, through Mexico and down the west coast of South America (2).
Or any other valid point.   4

(b) 1 mark for a valid point.
2 marks for a developed point.
If no area is named, mark out of 5.
Answers may include:
In the area of eastern Turkey in October 2011, immediate aid was required to help hundreds of trapped people (1). Emergency rescue teams were brought in from all over the world, using sniffer dogs and infra-red cameras to locate people trapped in collapsed buildings (1). Tents and blankets were given to people made homeless (1). Bottled water was given to people, as water supplies had been cut off (1). Medical help was given to injured people (1). Money was given by other countries to help rebuild schools, hospitals and clinics (1). Foreign governments and the UN gave aid to help rebuild the infrastructure – roads, water supplies, electricity supplies and bridges (1). Farmers were supplied with seeds and machinery to help grow crops for the future (1).
Or any other valid point.   6

12 (a) 1 mark for a valid point.
2 marks for a developed point.
Answers may include:
The country's main export consists of machinery and electrical goods (64%), and its main import is also machinery and electrical goods (32%) (2). It imports and exports the same percentage of manufactured goods, with 11% respectively (1). Similarly, the amount of exports and imports of chemicals is almost the same at 6% and 7% respectively (1). The only raw materials imported include oil (18%) and timber (3%) (1). It does not export any raw materials (1). The second highest export consists of vehicles at 16% (1).
Or any other valid point.   4

(b) 1 mark for a valid point.
2 marks for a developed point.
Answers may include:
Developing countries are often overdependent on raw materials for export (1). Many are overdependent on just one or two raw materials. For example, Zambia is dependent on copper – 87% of its exports – and Ghana depends on cocoa for 80% of its exports (2). If the prices of these commodities fall, then the countries will receive much less income (1). This means less money to buy imports or to pay for public services such as schools and hospitals (2). Workers will receive less money in their pay (1). Standards of living will fall (1). The country may have to borrow money from richer countries (1). The country's debt will rise, and its economy may fail (1).
Or any other valid point.   6

13 (a) 1 mark for a valid point.
2 marks for a developed point.
Answers may include:
Apart from France and the United Kingdom, the general trend is that of an increase in arrivals during the period 2008 to 2010 (1). The number of arrivals for Malaysia and Mexico has remained almost the same during this period (1). The largest increases took place in Spain, the United States, China and Italy (1). The combined total of arrivals in European countries vastly exceeds the number of arrivals in the USA and China (1).
Or any other valid point.   4

(b) 1 mark per valid point.
2 marks for a developed point.
If no named country, mark out of 5.
Answers may include:
Eco-tourism, in countries such as South Africa, helps people by bringing more money into the economy (1). Eco-tourism helps to create employment in a range of activities including hotels, farming, transport and retailing (2). Eco-tourism helps to protect the culture and traditions of native populations (1). Money from tourism helps to improve the local and natural environment by ensuring that beaches remain clean, historic buildings and sites are maintained, and wildlife and safari parks are set up in, for example, many African countries (2). Eco-tourism is aimed at encouraging awareness of the ecological damage that tourist development often entails (1). Eco-tourists are encouraged to visit certain areas to gain an understanding of the lifestyles and culture of local people and an appreciation of the natural environment and local ecosystems (2).
Or any other valid point.   6

14 (a) 1 mark for a valid point.
2 marks for a developed point.
Answers may include:
The number of cases in the provinces has dropped dramatically from about 63000 cases in 2000 to about 6000 in 2010 (1). The largest decrease has occurred in KwaZulu Natal Province, with a drop of over 40000 in the period shown (1). In 2010, there were no cases of malaria in KZN province (1). The number of cases in the other two provinces decreased from about 21000 in 2000 to about 6000 in 2010 (1).
Or any other valid point.   4

(b) 1 mark for a valid point.
   2 marks for a developed point.
   If Malaria chosen.
   Methods include:
      Draining areas with stagnant water, for example
      swamps, to destroy the breeding grounds of
      mosquitoes (2); using insecticides such as Malathion
      to kill mosquitoes (1); using nets to protect people
      from mosquito bites while they are sleeping (1);
      using drugs such as chloroquine to control the
      disease (1); setting up village health centres to
      provide information and education programmes (1);
      releasing water from dams to drown mosquito larvae
      (1); introducing small fish in paddy fields to eat the
      larvae (1); adding mustard seeds to water to drag
      larvae below the surface and drown them (2).
   Or any other valid point.   6

Acknowledgements

Permission has been sought from all relevant copyright holders and Hodder Gibson is grateful for the use of the following:

Image © gary yim/Shutterstock.com (Model Paper 1 page 6);

The diagram 'Land Use in the Rainforest' © Rhett A. Butler/Mongabay.com (Model Paper 1 page 10);

The diagram 'Tourist arrivals 1990–2012' adapted from: http://en.ria.ru/infographics/20120928/176277103.html?id= © RIA Novosti/UNWTO (United Nations World Tourism Organization) (Model Paper 1 page 13);

Image © Oliver Hoffmann/Shutterstock.com (Model Paper 1 page 14);

Image © Africa Studio/Shutterstock.com (Model Paper 1 page 14);

The diagram average 'Global Temperatures 1880–2012'. Data source: NASA Goddard Institute for Space Studies. Image credit: NASA Earth Observatory, Robert Simmon (Model Paper 2 page 9);

The diagram 'Distribution of Tropical Storms' © Caitlin.m/Creative Commons (CC BY-SA 3.0) http://creativecommons.org/licenses/by-sa/3.0/deed.en (Model Paper 2 page 11);

The diagram 'Cholera outbreaks 2010–2011', taken from http://www.soziologie-etc.com/med/ziv-u-korr/medizinkartell/02-d/008-weltkarte-cholera-ausbrueche2010-2011.gif. Data Source: World Health Organization. Map Production: Public Health Information and Geographic Information Systems (GIS) World Health Organization © Geneva, World Health Organization, 2012 (Model Paper 2 page 14);

The diagram 'World Population Growth (projected)', taken from United Nations (U.N.) Population Division, Long-Range World Population Projections: Two Centuries of World Population Growth, 1950-2150 (U.N., New York, 1992), p 22 (Model Paper 3 page 8);

The diagram 'Distribution of Earthquakes, October 2012'. Image courtesy USGS (public domain) (Model Paper 3 page 12);

Ordnance Survey maps © Crown Copyright 2013. Ordnance Survey 100047450.

Hodder Gibson would like to thank SQA for use of any past exam questions that may have been used in model papers, whether amended or in original form.